Basic Math & Pre-Algebra For Dummies

2nd Edition

by Mark Zegarelli

Basic Math & Pre-Algebra For Dummies®, 2nd Edition

Published by: **John Wiley & Sons, Inc.,** 111 River Street, Hoboken, NJ 07030-5774, www.wiley.com

For general information on our other products and services, please contact our Customer Care Department within the U.S. at 877-762-2974, outside the U.S. at 317-572-3993, or fax 317-572-4002. For technical support, please visit www.wiley.com/techsupport.

Wiley publishes in a variety of print and electronic formats and by print-on-demand. Some material included with standard print versions of this book may not be included in e-books or in print-on-demand. If this book refers to media such as a CD or DVD that is not included in the version you purchased, you may download this material at http://booksupport.wiley.com. For more information about Wiley products, visit www.wiley.com.

Library of Congress Control Number: 2013952434

ISBN 978-1-118-79198-1 (pbk); ISBN 978-1-118-79199-8 (ebk); ISBN 978-1-118-79205-6 (ebk)

Manufactured in the United States of America

10 9 8 7 6 5 4 3 2 1

Contents at a Glance

Table of Contents

Introduction

Once upon a time, you loved numbers. This isn't the first line of a fairy tale. Once upon a time, you really did love numbers. Remember?

Maybe you were 3 years old and your grandparents were visiting. You sat next to them on the couch and recited the numbers from 1 to 10. Grandma and Grandpa were proud of you and — be honest — you were proud of yourself, too. Or maybe you were 5 and discovering how to write numbers, trying hard not to print your 6 and 7 backward.

Learning was fun. *Numbers* were fun. So what happened? Maybe the trouble started with long division. Or sorting out how to change fractions to decimals. Could it have been figuring out how to add 8 percent sales tax to the cost of a purchase? Reading a graph? Converting miles to kilometers? Trying to find that most dreaded value of x? Wherever it started, you began to suspect that math didn't like you — and you didn't like math very much, either.

Why do people often enter preschool excited about learning how to count and leave high school as young adults convinced that they can't do math? The answer to this question would probably take 20 books this size, but solving the problem can begin right here.

I humbly ask you to put aside any doubts. Remember, just for a moment, an innocent time — a time before math-inspired panic attacks or, at best, induced irresistible drowsiness. In this book, I take you from an understanding of the basics to the place where you're ready to enter any algebra class and succeed.

About This Book

Somewhere along the road from counting to algebra, most people experience the Great Math Breakdown. This feels something like when your car begins smoking and sputtering on a 110°F highway somewhere between Noplace and Not Much Else.

Please consider this book your personal roadside helper, and think of me as your friendly math mechanic (only much cheaper!). Stranded on the interstate, you may feel frustrated by circumstances and betrayed by your vehicle, but for the guy holding the toolbox, it's all in a day's work. The tools for fixing the problem are in this book.

Not only does this book help you with the basics of math, but it also helps you get past any aversion you may feel toward math in general. I've broken down the concepts into easy-to-understand sections. And because *Basic Math & Pre-Algebra For Dummies* is a reference book, you don't have to read the chapters or sections in order — you can look over only what you need. So feel free to jump around. Whenever I cover a topic that requires information from earlier in the book, I refer you to that section or chapter, in case you want to refresh yourself on the basics.

Here are two pieces of advice I give all the time — remember them as you work your way through the concepts in this book:

- **Take frequent breaks.** Every 20 to 30 minutes, stand up and push in your chair. Then feed the cat, do the dishes, take a walk, juggle tennis balls, try on last year's Halloween costume — do *something* to distract yourself for a few minutes. You'll come back to your books more productive than if you just sat there hour after hour with your eyes glazing over.
- **After you've read through an example and think you understand it, copy the problem, close the book, and try to work it through.** If you get stuck, steal a quick look — but later, try that same example again to see whether you can get through it without opening the book. (Remember that, on any tests you're preparing for, peeking is probably not allowed!)

Although every author secretly (or not-so-secretly) believes that each word he pens is pure gold, you don't have to read every word in this book unless you really want to. Feel free to skip over sidebars (those shaded gray boxes) where I go off on a tangent — unless you find tangents interesting, of course. Paragraphs labeled with the Technical Stuff icon are also nonessential.

Foolish Assumptions

If you're planning to read this book, you likely fall into one of these categories:

- A student who wants a solid understanding of the basics of math for a class or test you're taking
- An adult who wants to improve skills in arithmetic, fractions, decimals, percentages, weights and measures, geometry, algebra, and so on for when you have to use math in the real world
- Someone who wants a refresher so you can help another person understand math

My only assumption about your skill level is that you can add, subtract, multiply, and divide. So to find out whether you're ready for this book, take this simple test:

$$5 + 6 = \_\_\_$$
$$10 - 7 = \_\_\_$$
$$3 \times 5 = \_\_\_$$
$$20 \div 4 = \_\_\_$$

If you can answer these four questions, you're ready to begin.

Icons Used in This Book

Throughout the book, I use four icons to highlight what's hot and what's not:

This icon points out key ideas that you need to know. Make sure you understand before reading on! Remember this info even after you close the book.

Tips are helpful hints that show you the quick and easy way to get things done. Try them out, especially if you're taking a math course.

Warnings flag common errors that you want to avoid. Get clear about where these little traps are hiding so you don't fall in.

This icon points out interesting trivia that you can read or skip over as you like.

Beyond the Book

In addition to the material in the print or e-book you're reading right now, remember that (as they say on those late-night infomercials) "There's much, much more!" Be sure to check out the free Cheat Sheet at `www.Dummies.com/cheatsheet/basicmathanndprealgebra` for a set of quick reference notes on converting between English and metric measurement units; using the order of operations (also called order of precedence); working with the commutative, associative, and distributive properties; converting among fractions, decimals, and percents; and lots, lots more.

In addition, `www.Dummies.com/webextras/basicmathandprealgebra` also contains a set of related material on topics like how to use factor trees to find the greatest common factor (GCF) of two or more numbers; how to use the percent circle, a helpful tool for solving percent problems; how to calculate the probability of getting certain rolls in the casino game of craps, and more.

And remember that in math, practice makes perfect. The *Basic Math & Pre-Algebra Workbook For Dummies* includes hundreds of practice problems, each group with a brief explanation to help you get started. And if that's not enough practice, *1,001 Practice Problems in Basic Math & Pre-Algebra For Dummies* provides lots more. Check them out!

Where to Go from Here

You can use this book in a few ways. If you're reading this book without immediate time pressure from a test or homework assignment, you can certainly start at the beginning and keep going to the end. The advantage to this method is that you realize how much math you *do* know — the first few chapters go very quickly. You gain a lot of confidence, as well as some practical knowledge that can help you later, because the early chapters also set you up to understand what follows.

If your time is limited — especially if you're taking a math course and you're looking for help with your homework or an upcoming test — skip directly to the topic you're studying. Wherever you open the book, you can find a clear explanation of the topic at hand, as well as a variety of hints and tricks. Read through the examples and try to do them yourself, or use them as templates to help you with assigned problems. Here's a short list of topics that tend to back students up:

- Negative numbers (Chapter 4)
- Order of operations (Chapter 5)
- Word problems (Chapters 6, 13, 18, and 23)
- Factoring of numbers (Chapter 8)
- Fractions (Chapters 9 and 10)

Generally, any time you spend building these five skills is like money in the bank as you proceed in math, so you may want to visit these sections several times.

Part I

Getting Started with Basic Math and Pre-Algebra

getting started
with
basic math
and
pre-algebra

Visit `www.dummies.com` for great Dummies content online.

In this part...

- See how the number system was invented and how it works
- Identify four important sets of numbers: counting numbers, integers, rational numbers, and real numbers
- Use place value to write numbers of any size
- Round numbers to make calculating quicker
- Work with the Big Four operations: adding, subtracting, multiplying, and dividing

Chapter 1

Playing the Numbers Game

In This Chapter

- Finding out how numbers were invented
- Looking at a few familiar number sequences
- Examining the number line
- Understanding four important sets of numbers

One useful characteristic about numbers is that they're *conceptual,* which means that, in an important sense, they're all in your head. (This fact probably won't get you out of having to know about them, though — nice try!)

For example, you can picture three of anything: three cats, three baseballs, three cannibals, three planets. But just try to picture the concept of three all by itself, and you find it's impossible. Oh, sure, you can picture the numeral 3, but the *threeness* itself — much like love or beauty or honor — is beyond direct understanding. But when you understand the *concept* of three (or four, or a million), you have access to an incredibly powerful system for understanding the world: mathematics.

In this chapter, I give you a brief history of how numbers came into being. I discuss a few common *number sequences* and show you how these connect with simple math *operations* like addition, subtraction, multiplication, and division.

After that, I describe how some of these ideas come together with a simple yet powerful tool: the *number line.* I discuss how numbers are arranged on the number line, and I also show you how to use the number line as a calculator for simple arithmetic. Finally, I describe how the *counting numbers* (1, 2, 3, ...) sparked the invention of more unusual types of numbers, such as *negative numbers, fractions,* and *irrational numbers.* I also show you how these *sets of numbers* are *nested* — that is, how one set of numbers fits inside another, which fits inside another.

Inventing Numbers

Historians believe that the first number systems came into being at the same time as agriculture and commerce. Before that, people in prehistoric, hunter-gatherer societies were pretty much content to identify bunches of things as "a lot" or "a little."

But as farming developed and trade between communities began, more precision was needed. So people began using stones, clay tokens, and similar objects to keep track of their goats, sheep, oil, grain, or whatever commodity they had. They exchanged these tokens for the objects they represented in a one-to-one exchange.

Eventually, traders realized that they could draw pictures instead of using tokens. Those pictures evolved into tally marks and, in time, into more complex systems. Whether they realized it or not, their attempts to keep track of commodities led these early humans to invent something entirely new: *numbers*.

Throughout the ages, the Babylonians, Egyptians, Greeks, Romans, Mayans, Arabs, and Chinese (to name just a few) all developed their own systems of writing numbers.

Although Roman numerals gained wide currency as the Roman Empire expanded throughout Europe and parts of Asia and Africa, the more advanced system that the Arabs invented turned out to be more useful. Our own number system, the Hindu–Arabic numbers (also called decimal numbers), is closely derived from these early Arabic numbers.

Understanding Number Sequences

Although humans invented numbers for counting commodities, as I explain in the preceding section, they soon put them to use in a wide range of applications. Numbers were useful for measuring distances, counting money, amassing an army, levying taxes, building pyramids, and lots more.

But beyond their many uses for understanding the external world, numbers have an internal order all their own. So numbers are not only an *invention,* but equally a *discovery:* a landscape that seems to exist independently, with its own structure, mysteries, and even perils.

One path into this new and often strange world is the *number sequence:* an arrangement of numbers according to a rule. In the following sections, I introduce you to a variety of number sequences that are useful for making sense of numbers.

Evening the odds

One of the first facts you probably heard about numbers is that all of them are either even or odd. For example, you can split an even number of marbles *evenly* into two equal piles. But when you try to divide an odd number of marbles the same way, you always have one *odd,* leftover marble. Here are the first few even numbers:

2 4 6 8 10 12 14 16 ...

You can easily keep the sequence of even numbers going as long as you like. Starting with the number 2, keep adding 2 to get the next number.

Similarly, here are the first few odd numbers:

1 3 5 7 9 11 13 15 ...

The sequence of odd numbers is just as simple to generate. Starting with the number 1, keep adding 2 to get the next number.

Patterns of even or odd numbers are the simplest number patterns around, which is why kids often figure out the difference between even and odd numbers soon after learning to count.

Counting by threes, fours, fives, and so on

When you get used to the concept of counting by numbers greater than 1, you can run with it. For example, here's what counting by threes, fours, and fives looks like:

Threes: 3 6 9 12 15 18 21 24 ...

Fours: 4 8 12 16 20 24 28 32 ...

Fives: 5 10 15 20 25 30 35 40 ...

Counting by a given number is a good way to begin learning the multiplication table for that number, especially for the numbers you're kind of sketchy on. (In general, people seem to have the most trouble multiplying by 7, but 8 and 9 are also unpopular.) In Chapter 3, I show you a few tricks for memorizing the multiplication table once and for all.

These types of sequences are also useful for understanding factors and multiples, which you get a look at in Chapter 8.

Getting square with square numbers

When you study math, sooner or later, you probably want to use visual aids to help you see what numbers are telling you. (Later in this book, I show you how one picture can be worth a thousand numbers when I discuss geometry in Chapter 16 and graphing in Chapter 17.)

The tastiest visual aids you'll ever find are those little square cheese-flavored crackers. (You probably have a box sitting somewhere in the pantry. If not, saltine crackers or any other square food works just as well.) Shake a bunch out of a box and place the little squares together to make bigger squares. Figure 1-1 shows the first few.

1

1	2
3	4

1	2	3
4	5	6
7	8	9

1	2	3	4
5	6	7	8
9	10	11	12
13	14	15	16

1	2	3	4	5
6	7	8	9	10
11	12	13	14	15
16	17	18	19	20
21	22	23	24	25

Figure 1-1: Square numbers.

Illustration by Wiley, Composition Services Graphics

Voilà! The square numbers:

1 4 9 16 25 36 49 64 ...

You get a *square number* by multiplying a number by itself, so knowing the square numbers is another handy way to remember part of the multiplication table. Although you probably remember without help that $2 \times 2 = 4$ you may be sketchy on some of the higher numbers, such as $7 \times 7 = 49$. Knowing the square numbers gives you another way to etch that multiplication table forever into your brain, as I show you in Chapter 3.

Square numbers are also a great first step on the way to understanding exponents, which I introduce later in this chapter and explain in more detail in Chapter 4.

Composing yourself with composite numbers

Some numbers can be placed in rectangular patterns. Mathematicians probably should call numbers like these "rectangular numbers," but instead they chose the term *composite numbers*. For example, 12 is a composite number because you can place 12 objects in rectangles of two different shapes, as in Figure 1-2.

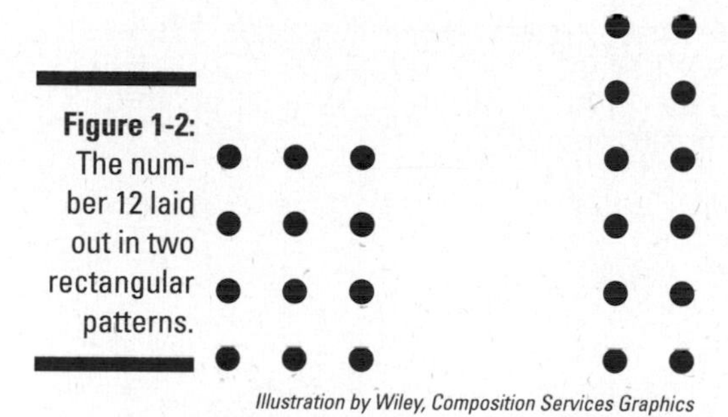

Figure 1-2: The number 12 laid out in two rectangular patterns.

Illustration by Wiley, Composition Services Graphics

As with square numbers, arranging numbers in visual patterns like this tells you something about how multiplication works. In this case, by counting the sides of both rectangles, you find out the following:

$3 \times 4 = 12$

$2 \times 6 = 12$

Similarly, other numbers such as 8 and 15 can also be arranged in rectangles, as in Figure 1-3.

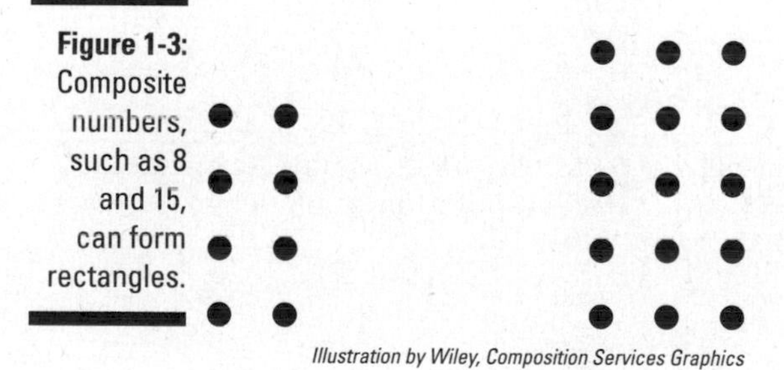

Figure 1-3: Composite numbers, such as 8 and 15, can form rectangles.

Illustration by Wiley, Composition Services Graphics

As you can see, both these numbers are quite happy being placed in boxes with at least two rows and two columns. And these visual patterns show this:

$2 \times 4 = 8$

$3 \times 5 = 15$

The word *composite* means that these numbers are *composed of* smaller numbers. For example, the number 15 is composed of 3 and 5 — that is, when you multiply these two smaller numbers, you get 15. Here are all the composite numbers from 1 to 16:

4 6 8 9 10 12 14 15 16

Notice that all the square numbers (see "Getting square with square numbers") also count as composite numbers because you can arrange them in boxes with at least two rows and two columns. Additionally, a lot of other nonsquare numbers are also composite numbers.

Stepping out of the box with prime numbers

Some numbers are stubborn. Like certain people you may know, these numbers — called *prime numbers* — resist being placed in any sort of a box. Look at how Figure 1-4 depicts the number 13, for example.

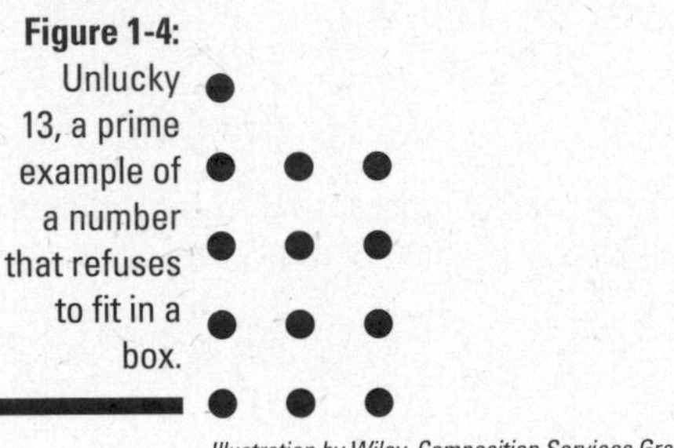

Figure 1-4: Unlucky 13, a prime example of a number that refuses to fit in a box.

Illustration by Wiley, Composition Services Graphics

Try as you may, you just can't make a rectangle out of 13 objects. (That fact may be one reason the number 13 got a bad reputation as unlucky.) Here are all the prime numbers less than 20:

2 3 5 7 11 13 17 19

As you can see, the list of prime numbers fills the gaps left by the composite numbers (see the preceding section). Therefore, every counting number is either prime or composite. The only exception is the number 1, which is neither prime nor composite. In Chapter 8, I give you a lot more information about prime numbers and show you how to *decompose* a number — that is, break down a composite number into its prime factors.

Multiplying quickly with exponents

Here's an old question whose answer may surprise you: Suppose you took a job that paid you just 1 penny the first day, 2 pennies the second day, 4 pennies the third day, and so on, doubling the amount every day, like this:

1 2 4 8 16 32 64 128 256 512 ...

As you can see, in the first ten days of work, you would've earned a little more than $10 (actually, $10.23 — but who's counting?). How much would you earn in 30 days? Your answer may well be, "I wouldn't take a lousy job like that in the first place." At first glance, this looks like a good answer, but here's a glimpse at your second ten days' earnings:

... 1,024 2,048 4,096 8,192 16,384 32,768 65,536 131,072 262,144 524,288 ...

By the end of the second 10 days, your total earnings would be over $10,000. And by the end of 30 days, your earnings would top out around $10,000,000! How does this happen? Through the magic of exponents (also called *powers*). Each new number in the sequence is obtained by multiplying the previous number by 2:

$$2^1 = 2$$

$$2^2 = 2 \times 2 = 4$$

$$2^3 = 2 \times 2 \times 2 = 8$$

$$2^4 = 2 \times 2 \times 2 \times 2 = 16$$

As you can see, the notation 2^4 means *multiply 2 by itself 4 times.*

You can use exponents on numbers other than 2. Here's another sequence you may be familiar with:

1 10 100 1,000 10,000 100,000 1,000,000 ...

In this sequence, every number is 10 times greater than the number before it. You can also generate these numbers using exponents:

$$10^1 = 10$$

$$10^2 = 10 \times 10 = 100$$

$$10^3 = 10 \times 10 \times 10 = 1,000$$

$$10^4 = 10 \times 10 \times 10 \times 10 = 10,000$$

This sequence is important for defining *place value,* the basis of the decimal number system, which I discuss in Chapter 2. It also shows up when I discuss decimals in Chapter 11 and scientific notation in Chapter 15. You find out more about exponents in Chapter 5.

Looking at the Number Line

As kids outgrow counting on their fingers (and use them only when trying to remember the names of all seven dwarfs), teachers often substitute a picture of the first ten numbers in order, like the one in Figure 1-5.

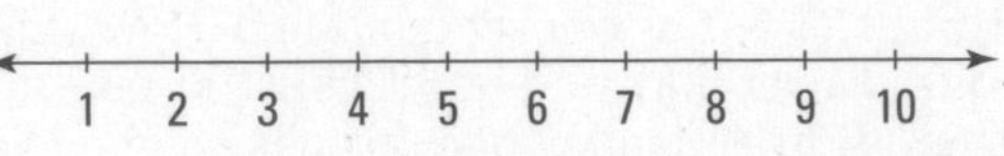

Figure 1-5: Basic number line.

This way of organizing numbers is called the *number line*. People often see their first number line — usually made of brightly colored construction paper — pasted above the blackboard in school. The basic number line provides a visual image of the *counting numbers* (also called the *natural numbers*), the numbers greater than 0. You can use it to show how numbers get bigger in one direction and smaller in the other.

In this section, I show you how to use the number line to understand a few basic but important ideas about numbers.

Adding and subtracting on the number line

You can use the number line to demonstrate simple addition and subtraction. These first steps in math become a lot more concrete with a visual aid. Here's the main point to remember:

- As you go *right,* the numbers go *up,* which is *addition* (+).
- As you go *left,* the numbers go *down,* which is *subtraction* (–).

For example, 2 + 3 means you *start at* 2 and *jump up* 3 spaces to 5, as Figure 1-6 illustrates.

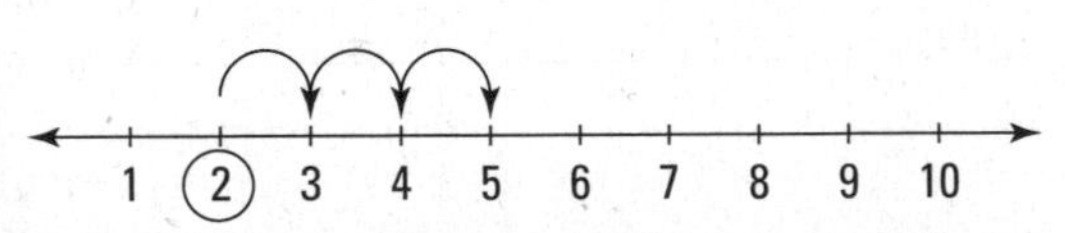

Figure 1-6: Moving through the number line from left to right.

As another example, 6 – 4 means *start* at 6 and *jump down* 4 spaces to 2. That is, 6 – 4 = 2. See Figure 1-7.

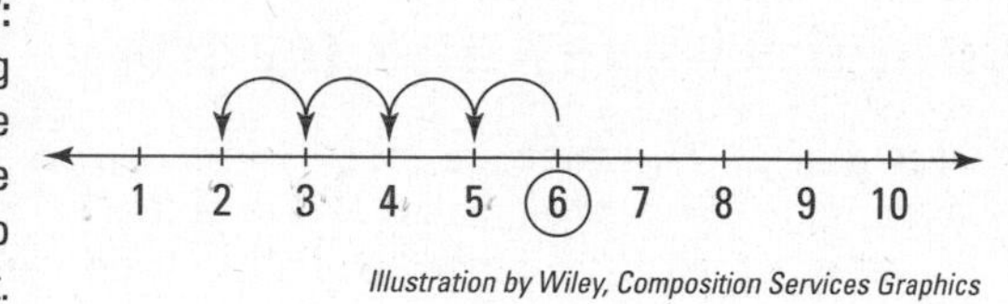

Figure 1-7: Moving through the number line from right to left.

You can use these simple up and down rules repeatedly to solve a longer string of added and subtracted numbers. For example, $3 + 1 - 2 + 4 - 3 - 2$ means 3, *up* 1, *down* 2, *up* 4, *down* 3, and *down* 2. In this case, the number line shows you that $3 + 1 - 2 + 4 - 3 - 2 = 1$.

I discuss addition and subtraction in greater detail in Chapter 3.

Getting a handle on nothing, or zero

An important addition to the number line is the number 0, which means *nothing, zilch, nada.* Step back a moment and consider the bizarre concept of nothing. For one thing — as more than one philosopher has pointed out — by definition, *nothing* doesn't exist! Yet we routinely label it with the number 0, as in Figure 1-8.

Actually, mathematicians have an even more precise labeling of *nothing* than zero. It's called the *empty* set, which is sort of the mathematical version of a box containing nothing. I introduce this concept, plus a little basic set theory, in Chapter 20.

Nothing sure is a heavy trip to lay on little kids, but they don't seem to mind. They understand quickly that when you have three toy trucks and someone

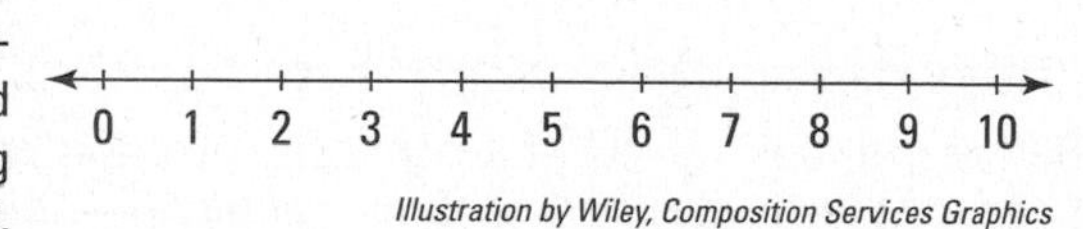

Figure 1-8: The number line starting at 0 and continuing with 1, 2, 3, ... 10.

Infinity: Imagining a never-ending story

The arrows at the ends of the number line point onward to a place called *infinity,* which isn't really a place at all — just the idea of *foreverness* because the numbers go on forever. But what about a million billion trillion quadrillion — do the numbers go even higher than that? The answer is yes, because for any number you name, you can add 1 to it.

The wacky symbol ∞ represents infinity. Remember, though, that ∞ isn't really a number but the *idea* that the numbers go on forever.

Because ∞ isn't a number, you can't technically add the number 1 to it, any more than you can add the number 1 to a cup of coffee or your Aunt Louise. But even if you could, $\infty + 1$ would equal ∞.

else takes away all three of them, you're left with zero trucks. That is, 3 – 3 = 0. Or, placing this on the number line, 3 – 3 means start at 3 and go down 3, as in Figure 1-9.

In Chapter 2, I show you the importance of 0 as a *placeholder* in numbers and discuss how you can attach *leading zeros* to a number without changing its value.

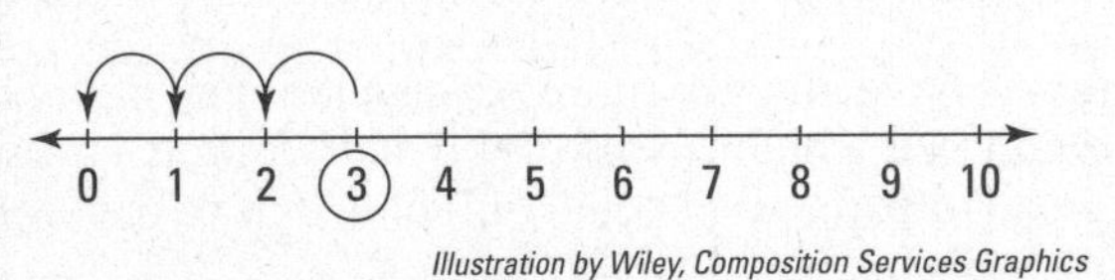

Figure 1-9: Starting at 3 and moving down three.

Illustration by Wiley, Composition Services Graphics

Taking a negative turn: Negative numbers

When people first find out about subtraction, they often hear that you can't take away more than you have. For example, if you have four pencils, you can take away one, two, three, or even all four of them, but you can't take away more than that.

It isn't long, though, before you find out what any credit card holder knows only too well: You can, indeed, take away more than you have — the result is a *negative number.* For example, if you have \$4 and you owe your friend \$7, you're \$3 in debt. That is, 4 – 7 = –3. The minus sign in front of the 3 means that the number of dollars you have is three less than 0. Figure 1-10 shows how you place negative whole numbers on the number line.

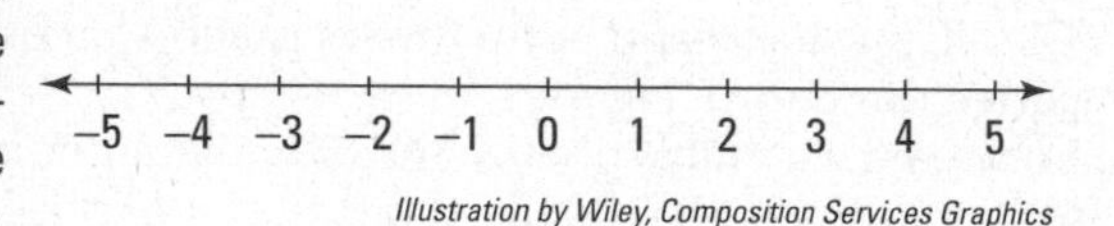

Figure 1-10: Negative whole numbers on the number line.

Adding and subtracting on the number line works pretty much the same with negative numbers as with positive numbers. For example, Figure 1-11 shows how to subtract 4 – 7 on the number line.

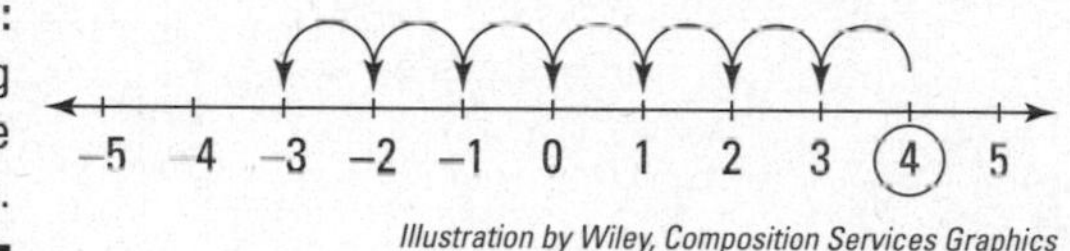

Figure 1-11: Subtracting 4 – 7 on the number line.

You find out all about working with negative numbers in Chapter 4.

Placing 0 and the negative counting numbers on the number line expands the set of counting numbers to the set of *integers*. I discuss the integers in further detail later in this chapter.

Multiplying the possibilities

Suppose you start at 0 and circle every other number on a number line, as in Figure 1-12. As you can see, all the even numbers are now circled. In other words, you've circled all the *multiples of two.* (You can find out more about multiples in Chapter 8.) You can now use this number line to multiply any number by two. For example, suppose you want to multiply 5 × 2. Just start at 0 and jump five circled spaces to the right.

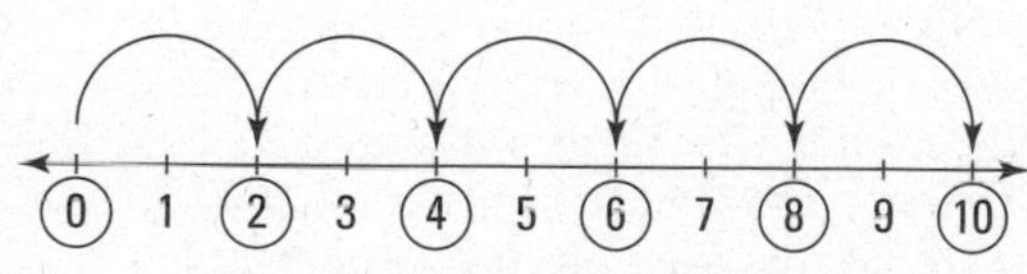

Figure 1-12: Multiplying 5 × 2 using the number line.

This number line shows you that $5 \times 2 = 10$.

Similarly, to multiply -3×2, start at 0 and jump three circled spaces to the left (that is, in the negative direction). Figure 1-13 shows you that $-3 \times 2 = -6$. What's more, you can now see why multiplying a negative number by a positive number always gives you a negative result. (I talk about multiplying by negative numbers in Chapter 4.)

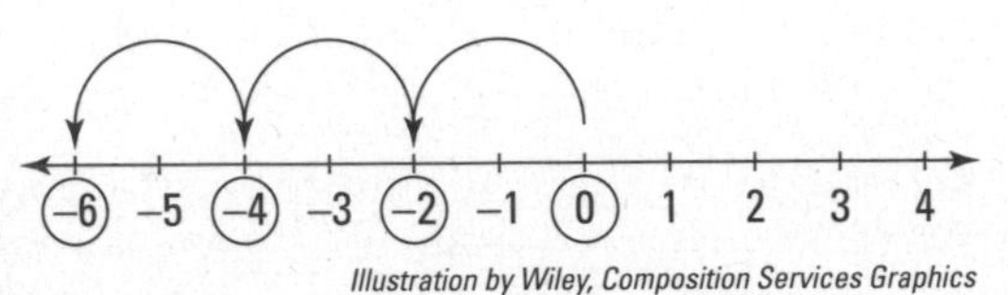

Figure 1-13: $3 \times 2 = -6$ as depicted on the number line.

Illustration by Wiley, Composition Services Graphics

Dividing things up

You can also use the number line to divide. For example, suppose you want to divide 6 by some other number. First, draw a number line that begins at 0 and ends at 6, as in Figure 1-14.

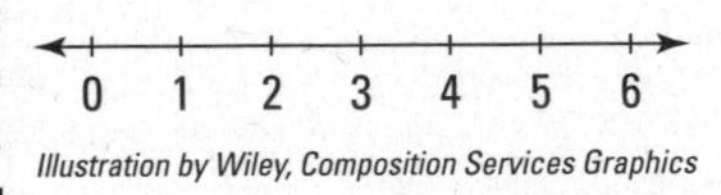

Figure 1-14: Number line from 0 to 6.

Illustration by Wiley, Composition Services Graphics

Now, to find the answer to $6 \div 2$, just split this number line into two equal parts, as in Figure 1-15. This split (or *division*) occurs at 3, showing you that $6 \div 2 = 3$.

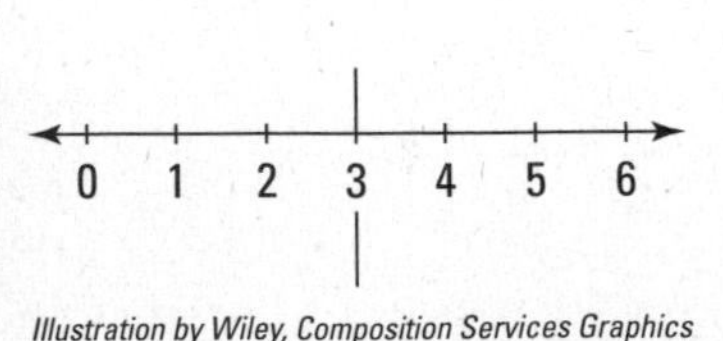

Figure 1-15: Getting the answer to $6 \div 2$ by splitting the number line.

Illustration by Wiley, Composition Services Graphics

Similarly, to divide 6 ÷ 3, split the same number line into three equal parts, as in Figure 1-16. This time you have two splits, so use the one closest to 0. This number line shows you that 6 ÷ 3 = 2.

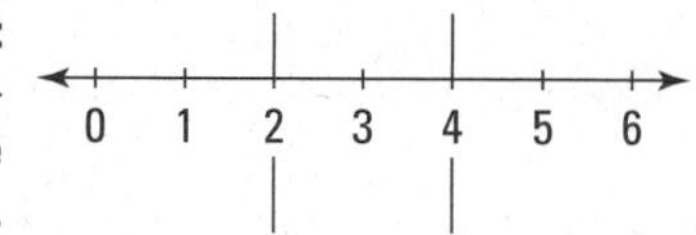

Figure 1-16: Dividing 6 ÷ 3 with the number line.

Illustration by Wiley, Composition Services Graphics

But suppose you want to use the number line to divide a small number by a larger number. For example, maybe you want to know the answer to 3 ÷ 4. Following the method I show you earlier, first draw a number line from 0 to 3. Then split it into four equal parts. Unfortunately, none of these splits has landed on a number. It's not a mistake — you just have to add some new numbers to the number line, as you can see in Figure 1-17.

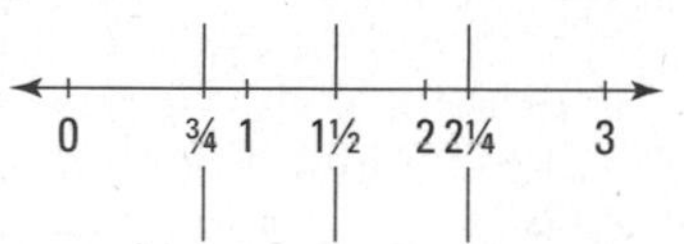

Figure 1-17: Fractions on the number line.

Illustration by Wiley, Composition Services Graphics

Welcome to the world of *fractions.* With the number line labeled properly, you can see that the split closest to 0 is $\frac{3}{4}$. This image tells you that $3 \div 4 = \frac{3}{4}$. The similarity of the expression 3 ÷ 4 and the fraction $\frac{3}{4}$ is no accident. Division and fractions are closely related. When you divide, you cut things up into equal parts, and fractions are often the result of this process. (I explain the connection between division and fractions in more detail in Chapters 9 and 10.)

Discovering the space in between: Fractions

Fractions help you fill in a lot of the spaces on the number line that fall between the counting numbers. For example, Figure 1-18 shows a close-up of a number line from 0 to 1.

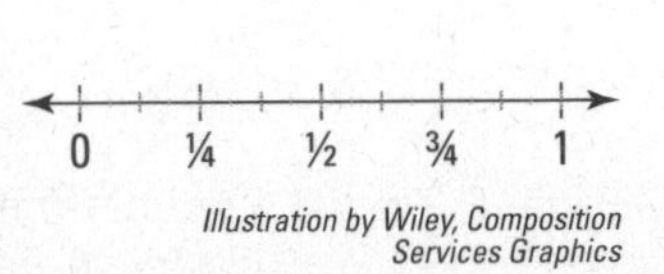

Illustration by Wiley, Composition Services Graphics

Figure 1-18: Number line depicting some fractions from 0 to 1.

This number line may remind you of a ruler or a tape measure, with a lot of tiny fractions filled in. In fact, rulers and tape measures really are portable number lines that allow carpenters, engineers, and savvy do-it-yourselfers to measure the length of objects with precision.

Adding fractions to the number line expands the set of integers to the set of *rational numbers.* I discuss the rational numbers in greater detail in Chapter 25.

In fact, no matter how small things get in the real world, you can always find a tiny fraction to approximate it as closely as you need. Between any two fractions on the number line, you can always find another fraction. Mathematicians call this trait the *density* of fractions on the real number line, and this type of density is a topic in a very advanced area of math called *real analysis.*

Four Important Sets of Numbers

In the preceding section, you see how the number line grows in both the positive and negative directions and fills in with a lot of numbers in between. In this section, I provide a quick tour of how numbers fit together as a set of nested systems, one inside the other.

When I talk about a set of numbers, I'm really just talking about a group of numbers. You can use the number line to deal with four important sets of numbers:

- **Counting numbers (also called natural numbers):** The set of numbers beginning 1, 2, 3, 4 ... and going on infinitely
- **Integers:** The set of counting numbers, zero, and negative counting numbers
- **Rational numbers:** The set of integers and fractions
- **Real numbers:** The set of rational and irrational numbers

The sets of counting numbers, integers, rational, and real numbers are nested, one inside another. This nesting of one set inside another is similar to the way that a city (for example, Boston) is inside a state (Massachusetts), which is inside a country (the United States), which is inside a continent (North America). The set of counting numbers is inside the set of integers, which is inside the set of rational numbers, which is inside the set of real numbers.

Counting on the counting numbers

The set of *counting numbers* is the set of numbers you first count with, starting with 1. Because they seem to arise naturally from observing the world, they're also called the *natural numbers:*

1 2 3 4 5 6 7 8 9...

The counting numbers are infinite, which means they go on forever.

When you add two counting numbers, the answer is always another counting number. Similarly, when you multiply two counting numbers, the answer is always a counting number. Another way of saying this is that the set of counting numbers is *closed* under both addition and multiplication.

Introducing integers

The set of *integers* arises when you try to subtract a larger number from a smaller one. For example, 4 – 6 = –2. The set of integers includes the following:

- The counting numbers
- Zero
- The negative counting numbers

Here's a partial list of the integers:

... –4 –3 –2 –1 0 1 2 3 4 ...

Like the counting numbers, the integers are closed under addition and multiplication. Similarly, when you subtract one integer from another, the answer is always an integer. That is, the integers are also closed under subtraction.

Staying rational

Here's the set of *rational numbers:*

- Integers
    - Counting numbers
    - Zero
    - Negative counting numbers
- Fractions

Like the integers, the rational numbers are closed under addition, subtraction, and multiplication. Furthermore, when you divide one rational number by another, the answer is always a rational number. Another way to say this is that the rational numbers are closed under division.

Getting real

Even if you filled in all the rational numbers, you'd still have points left unlabeled on the number line. These points are the irrational numbers.

An *irrational number* is a number that's neither a whole number nor a fraction. In fact, an irrational number can only be approximated as a *nonrepeating decimal.* In other words, no matter how many decimal places you write down, you can always write down more; furthermore, the digits in this decimal never become repetitive or fall into any pattern. (For more on repeating decimals, see Chapter 11.)

The most famous irrational number is π (you find out more about π when I discuss the geometry of circles in Chapter 17):

$$\pi = 3.14159265358979323846264338327950288419716939937510...$$

Together, the rational and irrational numbers make up the *real numbers,* which comprise every point on the number line. In this book, I don't spend too much time on irrational numbers, but just remember that they're there for future reference.

Chapter 2

It's All in the Fingers: Numbers and Digits

In This Chapter

- Understanding how place value turns digits into numbers
- Distinguishing whether zeros are important placeholders or meaningless leading zeros
- Reading and writing long numbers
- Understanding how to round numbers and estimate values

When you're counting, ten seems to be a natural stopping point — a nice, round number. The fact that our ten fingers match up so nicely with numbers may seem like a happy accident. But of course, it's no accident at all. Fingers were the first calculator that humans possessed. Our number system — Hindu-Arabic numbers — is based on the number ten because humans have 10 fingers instead of 8 or 12. In fact, the very word *digit* has two meanings: numerical symbol and finger.

In this chapter, I show you how place value turns digits into numbers. I also show you when 0 is an important placeholder in a number and why leading zeros don't change the value of a number. And I show you how to read and write long numbers. After that, I discuss two important skills: rounding numbers and estimating values.

Knowing Your Place Value

The number system you're most familiar with — Hindu-Arabic numbers — has ten familiar digits:

0 1 2 3 4 5 6 7 8 9

Telling the difference between numbers and digits

Sometimes people confuse numbers and digits. For the record, here's the difference:

- A digit is a single numerical symbol, from 0 to 9.
- A number is a string of one or more digits.

For example, 7 is both a digit and a number. In fact, it's a one-digit number. However, 15 is a string of two digits, so it's a number — a two-digit number. And 426 is a three-digit number. You get the idea.

In a sense, a digit is like a letter of the alphabet. By themselves, the uses of 26 letters, A through Z, are limited. (How much can you do with a single letter such as K or W?) Only when you begin using strings of letters as building blocks to spell words does the power of letters become apparent. Similarly, the ten digits, 0 through 9, have limited usefulness until you begin building strings of digits — that is, numbers.

Yet with only ten digits, you can express numbers as high as you care to go. In this section, I show you how it happens.

Counting to ten and beyond

The ten digits in our number system allow you to count from 0 to 9. All higher numbers are produced using place value. Place value assigns a digit a greater or lesser value, depending on where it appears in a number. Each place in a number is ten times greater than the place to its immediate right.

To understand how a whole number gets its value, suppose you write the number 45,019 all the way to the right in Table 2-1, one digit per cell, and add up the numbers you get.

Table 2-1 45,019 Displayed in a Place-Value Chart

Millions			*Thousands*			*Ones*		
Hundred Millions	Ten Millions	Millions	Hundred Thousands	Ten Thousands	Thousands	Hundreds	Tens	Ones
				4	5	0	1	9

You have 4 ten thousands, 5 thousands, 0 hundreds, 1 ten, and 9 ones. The chart shows you that the number breaks down as follows:

$$45{,}019 = 40{,}000 + 5{,}000 + 0 + 10 + 9$$

In this example, notice that the presence of the digit 0 in the hundreds place means that zero hundreds are added to the number.

Telling placeholders from leading zeros

Although the digit 0 adds no value to a number, it acts as a placeholder to keep the other digits in their proper places. For example, the number 5,001,000 breaks down into 5,000,000 + 1,000. Suppose, however, you decide to leave all the 0s out of the chart. Table 2-2 shows what you'd get.

Table 2-2 5,001,000 Displayed Incorrectly without Placeholding Zeros

Millions			*Thousands*			*Ones*		
Hundred Millions	Ten Millions	Mil-lions	Hundred Thousands	Ten Thou-sands	Thou-sands	Hun-dreds	Tens	Ones
							5	1

The chart tells you that 5,001,000 = 50 + 1. Clearly, this answer is wrong!

As a rule, when a 0 appears to the right of at least one digit other than 0, it's a placeholder. Placeholding zeros are important — always include them when you write a number. However, when a 0 appears to the left of every digit other than 0, it's a leading zero. Leading zeros serve no purpose in a number, so dropping them is customary. For example, place the number 003,040,070 on the chart (see Table 2-3).

Table 2-3 3,040,070 Displayed with Two Leading Zeros

Millions			*Thousands*			*Ones*		
Hundred Millions	Ten Millions	Mil-lions	Hundred Thousands	Ten Thou-sands	Thou-sands	Hun-dreds	Tens	Ones
0	0	3	0	4	0	0	7	0

The first two 0s in the number are leading zeros because they appear to the left of the 3. You can drop these 0s from the number, leaving you with 3,040,070. The remaining 0s are all to the right of the 3, so they're placeholders — be sure to write them in.

Reading long numbers

When you write a long number, you use commas to separate groups of three numbers. For example, here's about as long a number as you'll ever see:

234,845,021,349,230,467,304

Table 2-4 shows a larger version of the place-value chart.

Table 2-4 A Place-Value Chart Separated by Commas

Quintillions	*Quadrillions*	*Trillions*	*Billions*	*Millions*	*Thousands*	*Ones*
234	845	021	349	230	467	304

This version of the chart helps you read the number. Begin all the way to the left and read, "Two hundred thirty-four quintillion, eight hundred forty-five quadrillion, twenty-one trillion, three hundred forty-nine billion, two hundred thirty million, four hundred sixty-seven thousand, three hundred four."

When you read and write whole numbers, don't say the word *and*. In math, the word *and* means you have a decimal point. That's why, when you write a check, you save the word *and* for the number of cents, which is usually expressed as a decimal or sometimes as a fraction. (I discuss decimals in Chapter 11.)

Close Enough for Rock 'n' Roll: Rounding and Estimating

As numbers get longer, calculations become tedious, and you're more likely to make a mistake or just give up. When you're working with long numbers, simplifying your work by rounding numbers and estimating values is sometimes helpful.

When you round a number, you change some of its digits to placeholding zeros. And when you estimate a value, you work with rounded numbers to find an approximate answer to a problem. In this section, you build both skills.

Rounding numbers

Rounding numbers makes long numbers easier to work with. In this section, I show you how to round numbers to the nearest ten, hundred, thousand, and beyond.

Rounding numbers to the nearest ten

The simplest kind of rounding you can do is with two-digit numbers. When you round a two-digit number to the nearest ten, you simply bring it up or down to the nearest number that ends in 0. For example,

$$39 \rightarrow 40 \quad 51 \rightarrow 50 \quad 73 \rightarrow 70$$

Even though numbers ending in 5 are in the middle, always round them up to the next-highest number that ends in 0:

$$15 \rightarrow 20 \quad 35 \rightarrow 40 \quad 85 \rightarrow 90$$

Numbers in the upper 90s get rounded up to 100:

$$99 \rightarrow 100 \quad 95 \rightarrow 100 \quad 94 \rightarrow 90$$

When you know how to round a two-digit number, you can round just about any number. For example, to round most longer numbers to the nearest ten, just focus on the ones and tens digits:

$$734 \rightarrow 730 \quad 1{,}488 \rightarrow 1{,}490 \quad 12{,}345 \rightarrow 12{,}350$$

Occasionally, a small change to the ones and tens digits affects the other digits. (This situation is a lot like when the odometer in your car rolls a bunch of 9s over to 0s.) For example:

$$899 \rightarrow 900 \quad 1{,}097 \rightarrow 1{,}100 \quad 9{,}995 \rightarrow 10{,}000$$

Rounding numbers to the nearest hundred and beyond

To round numbers to the nearest hundred, thousand, or beyond, focus only on two digits: the digit in the place you're rounding to and the digit to its immediate right. Change all other digits to the right of these two digits to 0s. For example, suppose you want to round 642 to the nearest hundred. Focus on the hundreds digit (6) and the digit to its immediate right (4):

$$642$$

I've underlined these two digits. Now just round these two digits as if you were rounding to the nearest ten, and change the digit to the right of them to a 0:

$$\underline{64}2 \rightarrow 600$$

Here are a few more examples of rounding numbers to the nearest hundred:

$$7,\underline{89}1 \rightarrow 7,900 \quad 15,\underline{75}3 \rightarrow 15,800 \quad 99,\underline{96}1 \rightarrow 100,000$$

When rounding numbers to the nearest thousand, underline the thousands digit and the digit to its immediate right. Round the number by focusing only on the two underlined digits and, when you're done, change all digits to the right of these to 0s:

$$\underline{4,9}84 \rightarrow 5,000 \quad 7\underline{8,5}21 \rightarrow 79,000 \quad 1,09\underline{9,3}04 \rightarrow 1,099,000$$

Even when rounding to the nearest million, the same rules apply:

$$\underline{1,2}34,567 \rightarrow 1,000,000 \quad 7\underline{8,8}83,958 \rightarrow 79,000,000$$

Estimating value to make problems easier

When you know how to round numbers, you can use this skill in estimating values. Estimating saves you time by allowing you to avoid complicated computations and still get an approximate answer to a problem.

When you get an approximate answer, you don't use an equals sign; instead, you use this wavy symbol, which means *is approximately equal to:* ≈.

Suppose you want to add these numbers: 722 + 506 + 383 + 1,279 + 91 + 811. This computation is tedious, and you may make a mistake. But you can make the addition easier by first rounding all the numbers to the nearest hundred and then adding:

$$\approx 700 + 500 + 400 + 1,300 + 100 + 800 = 3,800$$

The approximate answer is 3,800. This answer isn't far off from the exact answer, which is 3,792.

Chapter 3

The Big Four: Addition, Subtraction, Multiplication, and Division

In This Chapter

- Reviewing addition
- Understanding subtraction
- Viewing multiplication as a fast way to do repeated addition
- Getting clear on division

When most folks think of math, the first thing that comes to mind is four little (or not-so-little) words: addition, subtraction, multiplication, and division. I call these operations the *Big Four* all through the book.

In this chapter, I introduce you (or reintroduce you) to these little gems. Although I assume you're already familiar with the Big Four, this chapter reviews these operations, taking you from what you may have missed to what you need to succeed as you move onward and upward in math.

Adding Things Up

Addition is the first operation you find out about, and it's almost everybody's favorite. It's simple, friendly, and straightforward. No matter how much you worry about math, you've probably never lost a minute of sleep over addition. Addition is all about bringing things together, which is a positive goal. For example, suppose you and I are standing in line to buy tickets for a movie. I have $25 and you have only $5. I could lord it over you and make you feel crummy that I can go to the movies and you can't. Or instead, you and

I can join forces, adding together my $25 and your $5 to make $30. Now, not only can we both see the movie, but we may even be able to buy some popcorn, too.

Addition uses only one sign — the plus sign (+): Your equation may read $2 + 3 = 5$, or $12 + 2 = 14$, or $27 + 44 = 71$, but the plus sign always means the same thing.

When you add two numbers together, those two numbers are called *addends,* and the result is called the *sum.* So in the first example, the addends are 2 and 3, and the sum is 5.

In line: Adding larger numbers in columns

When you want to add larger numbers, stack them on top of each other so that the ones digits line up in a column, the tens digits line up in another column, and so on. (Chapter 2 has the scoop on digits and place value.) Then add column by column, starting from the ones column on the right. Not surprisingly, this method is called *column addition*. Here's how you add 55 + 31 + 12. First add the ones column:

$$\begin{array}{r} 5\mathbf{5} \\ 3\mathbf{1} \\ +\ 1\mathbf{2} \\ \hline \mathbf{8} \end{array}$$

Next, move to the tens column:

$$\begin{array}{r} \mathbf{5}5 \\ \mathbf{3}1 \\ +\ \mathbf{1}2 \\ \hline \mathbf{9}8 \end{array}$$

This problem shows you that $55 + 31 + 12 = 98$.

Carry on: Dealing with two-digit answers

Sometimes when you're adding a column, the sum is a two-digit number. In that case, you need to write down the ones digit of that number and carry the tens digit over to the next column to the left — that is, write this digit above the column so you can add it with the rest of the numbers in that column. For example, suppose you want to add 376 + 49 + 18. In the ones column, 6 + 9 + 8 = 23, so write down the 3 and carry the 2 over to the top of the tens column:

$$\begin{array}{r} 2 \\ 376 \\ 49 \\ +\ 18 \\ \hline 3 \end{array}$$

Now continue by adding the tens column. In this column, 2 + 7 + 4 + 1 = 14, so write down the 4 and carry the 1 over to the top of the hundreds column:

$$\begin{array}{r} 12 \\ 376 \\ 49 \\ +\ 18 \\ \hline 43 \end{array}$$

Continue adding in the hundreds column:

$$\begin{array}{r} 12 \\ 376 \\ 49 \\ +\ 18 \\ \hline 443 \end{array}$$

This problem shows you that 376 + 49 + 18 = 443.

Take It Away: Subtracting

Subtraction is usually the second operation you discover, and it's not much harder than addition. Still, there's something negative about subtraction — it's all about who has more and who has less. Suppose you and I have been running on treadmills at the gym. I'm happy because I ran 3 miles, but then you start bragging that you ran 10 miles. You subtract and tell me that I should be very impressed that you ran 7 miles farther than I did. (But with an attitude like that, don't be surprised if you come back from the showers to find your running shoes filled with liquid soap!)

As with addition, subtraction has only one sign: the minus sign (–). You end up with equations such as 4 – 1 = 3, and 14 – 13 = 1, and 93 – 74 = 19.

When you subtract one number from another, the result is called the *difference*. This term makes sense when you think about it: When you subtract, you find the difference between a higher number and a lower one.

In subtraction, the first number is called the *minuend,* and the second number is called the *subtrahend.* But almost nobody ever remembers which is which, so when I talk about subtraction, I prefer to say *the first number* and *the second number.*

One of the first facts you probably heard about subtraction is that you can't take away more than you start with. In that case, the second number can't be larger than the first. And if the two numbers are the same, the result is always 0. For example, 3 – 3 = 0; 11 – 11 = 0; and 1,776 – 1,776 = 0. Later someone breaks the news that you *can* take away more than you have. When you do, though, you need to place a minus sign in front of the difference to show that you have a negative number, a number below 0:

$4 - 5 = -1$

$10 - 13 = -3$

$88 - 99 = -11$

When subtracting a larger number from a smaller number, remember the words *switch* and *negate:* You *switch* the order of the two numbers and do the subtraction as you normally would, but at the end, you *negate* the result by attaching a minus sign. For example, to find 10 – 13, you switch the order of these two numbers, giving you 13 – 10, which equals 3; then you negate this result to get –3. That's why 10 – 13 = –3.

The minus sign does double duty, so don't get confused. When you stick a minus sign between two numbers, it means the first number minus the second number. But when you attach it to the front of a number, it means that this number is a negative number.

Flip to Chapter 1 to see how negative numbers work on the number line. I also go into more detail on negative numbers and the Big Four operations in Chapter 4.

Columns and stacks: Subtracting larger numbers

To subtract larger numbers, stack one on top of the other as you do with addition. (For subtraction, however, don't stack more than two numbers —

put the larger number on top and the smaller one underneath it.) For example, suppose you want to subtract 386 – 54. To start, stack the two numbers and begin subtracting in the ones column: 6 – 4 = 2:

$$\begin{array}{r} 38\mathbf{6} \\ \underline{-5\mathbf{4}} \\ \mathbf{2} \end{array}$$

Next, move to the tens column and subtract 8 – 5 to get 3:

$$\begin{array}{r} 3\mathbf{8}6 \\ \underline{-\mathbf{5}4} \\ \mathbf{3}2 \end{array}$$

Finally, move to the hundreds column. This time, 3 – 0 = 3:

$$\begin{array}{r} \mathbf{3}86 \\ \underline{-54} \\ \mathbf{3}32 \end{array}$$

This problem shows you that 386 – 54 = 332.

Can you spare a ten? Borrowing to subtract

Sometimes the top digit in a column is smaller than the bottom digit in that column. In that case, you need to borrow from the next column to the left. Borrowing is a two-step process:

1. **Subtract 1 from the top number in the column directly to the left.**

   Cross out the number you're borrowing from, subtract 1, and write the answer above the number you crossed out.

2. **Add 10 to the top number in the column you were working in.**

For example, suppose you want to subtract 386 – 94. The first step is to subtract 4 from 6 in the ones column, which gives you 2:

$$\begin{array}{r} 38\mathbf{6} \\ \underline{-9\mathbf{4}} \\ \mathbf{2} \end{array}$$

When you move to the tens column, however, you find that you need to subtract 8 – 9. Because 8 is smaller than 9, you need to borrow from the hundreds column. First, cross out the 3 and replace it with a 2, because 3 – 1 = 2:

$$\begin{array}{r} \mathbf{2} \\ \not{3}86 \\ -94 \\ \hline 2 \end{array}$$

Next, place a 1 in front of the 8, changing it to an 18, because 8 + 10 = 18:

$$\begin{array}{r} 2 \\ \not{3}\mathbf{1}86 \\ -94 \\ \hline 2 \end{array}$$

Now you can subtract in the tens column: 18 – 9 = 9:

$$\begin{array}{r} 2\mathbf{18}6 \\ -\mathbf{9}4 \\ \hline \mathbf{9}2 \end{array}$$

The final step is simple: 2 – 0 = 2:

$$\begin{array}{r} \mathbf{2}186 \\ -94 \\ \hline \mathbf{2}92 \end{array}$$

Therefore, 386 – 94 = 292.

In some cases, the column directly to the left may not have anything to lend. Suppose, for instance, that you want to subtract 1,002 – 398. Beginning in the ones column, you find that you need to subtract 2 – 8. Because 2 is smaller than 8, you need to borrow from the next column to the left. But the digit in the tens column is a 0, so you can't borrow from there because the cupboard is bare, so to speak:

$$\begin{array}{r} 10\mathbf{02} \\ -39\mathbf{8} \\ \hline \end{array}$$

When borrowing from the next column isn't an option, you need to borrow from the nearest nonzero column to the left.

In this example, the column you need to borrow from is the thousands column. First, cross out the 1 and replace it with a 0. Then place a 1 in front of the 0 in the hundreds column:

0
~~1~~ **10** 0 2
−3 9 8

Now cross out the 10 and replace it with a 9. Place a 1 in front of the 0 in the tens column:

0 **9**
~~1~~ **~~10~~** **10**2
−3 98

Finally, cross out the 10 in the tens column and replace it with a 9. Then place a 1 in front of the 2:

0 9 **9**
~~1~~ ~~10~~ **~~10~~** **12**
−3 9 **8**

At last, you can begin subtracting in the ones column: 12 – 8 = 4:

0 9 9
~~1~~ ~~10~~ ~~10~~ **12**
−3 9 **8**
4

Then subtract in the tens column: 9 – 9 = 0:

0 9 **9**
~~1~~ ~~10~~ ~~10~~ 12
−3 **9** 8
0 4

Then subtract in the hundreds column: 9 – 3 = 9:

0 9 **9**
~~1~~ ~~10~~ ~~10~~ 12
−3 **9** 8
6 **0** 4

Because nothing is left in the thousands column, you don't need to subtract anything else. Therefore, 1,002 – 398 = 604.

Multiplying

Multiplication is often described as a sort of shorthand for repeated addition. For example,

4×3 means add 4 to itself 3 times: $4 + 4 + 4 = 12$

9×6 means add 9 to itself 6 times: $9 + 9 + 9 + 9 + 9 + 9 = 54$

100×2 means add 100 to itself 2 times: $100 + 100 = 200$

Although multiplication isn't as warm and fuzzy as addition, it's a great time-saver. For example, suppose you coach a Little League baseball team, and you've just won a game against the toughest team in the league. As a reward, you promised to buy three hot dogs for each of the nine players on the team. To find out how many hot dogs you need, you can add 3 together 9 times. Or you can save time by multiplying 3 times 9, which gives you 27. Therefore, you need 27 hot dogs (plus a whole lot of mustard and sauerkraut).

When you multiply two numbers, the two numbers that you're multiplying are called *factors,* and the result is the *product.*

In multiplication, the first number is also called the *multiplicand* and the second number is the *multiplier.* But almost nobody ever remembers — or uses — these words.

Signs of the times

When you're first introduced to multiplication, you use the times sign (×). As you move onward and upward on your math journey, you need to be aware of the conventions I discuss in the following sections.

The symbol · is sometimes used to replace the symbol ×. For example,

$$\begin{aligned} 4 \cdot 2 = 8 &\quad \text{means} \quad 4 \times 2 = 8 \\ 6 \cdot 7 = 42 &\quad \text{means} \quad 6 \times 7 = 42 \\ 53 \cdot 11 = 583 &\quad \text{means} \quad 53 \times 11 = 583 \end{aligned}$$

In Parts I through IV of this book, I stick to the tried-and-true symbol × for multiplication. Just be aware that the symbol · exists so that you won't be stumped if your teacher or textbook uses it.

In math beyond arithmetic, using parentheses without another operator stands for multiplication. The parentheses can enclose the first number, the second number, or both numbers. For example,

$$3(5) = 15 \quad \text{means} \quad 3 \times 5 = 15$$
$$(8)7 = 56 \quad \text{means} \quad 8 \times 7 = 56$$
$$(9)(10) = 90 \quad \text{means} \quad 9 \times 10 = 90$$

This switch makes sense when you stop to consider that the letter x, which is often used in algebra, looks a lot like the multiplication sign ×. So in this book, when I start using x in Part V, I also stop using × and begin using parentheses without another sign to indicate multiplication.

Memorizing the multiplication table

You may consider yourself among the multiplicationally challenged. That is, you consider being called upon to remember 9 × 7 a tad less appealing than being dropped from an airplane while clutching a parachute purchased from the trunk of some guy's car. If so, then this section is for you.

Looking at the old multiplication table

One glance at the old multiplication table, Table 3-1, reveals the problem. If you saw the movie *Amadeus,* you may recall that Mozart was criticized for writing music that had "too many notes." Well, in my humble opinion, the multiplication table has too many numbers.

Table 3-1 The Monstrous Standard Multiplication Table

	0	***1***	***2***	***3***	***4***	***5***	***6***	***7***	***8***	***9***
0	0	0	0	0	0	0	0	0	0	0
1	0	1	2	3	4	5	6	7	8	9
2	0	2	4	6	8	10	12	14	16	18
3	0	3	6	9	12	15	18	21	24	27
4	0	4	8	12	16	20	24	28	32	36
5	0	5	10	15	20	25	30	35	40	45

(continued)

Table 3-1 *(continued)*

	0	*1*	*2*	*3*	*4*	*5*	*6*	*7*	*8*	*9*
6	0	6	12	18	24	30	36	42	48	54
7	0	7	14	21	28	35	42	49	56	63
8	0	8	16	24	32	40	48	56	64	72
9	0	9	18	27	36	45	54	63	72	81

I don't like the multiplication table any more than you do. Just looking at it makes my eyes glaze over. With 100 numbers to memorize, no wonder so many folks just give up and carry a calculator.

Introducing the short multiplication table

If the multiplication table from Table 3-1 were smaller and a little more manageable, I'd like it a lot more. So here's my short multiplication table, in Table 3-2.

Table 3-2 The Short Multiplication Table

	3	*4*	*5*	*6*	*7*	*8*	*9*
3	9	12	15	18	21	24	27
4		16	20	24	28	32	36
5			25	30	35	40	45
6				36	42	48	54
7					49	56	63
8						64	72
9							81

As you can see, I've gotten rid of a bunch of numbers. In fact, I've reduced the table from 100 numbers to 28. I've also shaded 11 of the numbers I've kept.

Is just slashing and burning the sacred multiplication table wise? Is it even legal? Well, of course it is! After all, the table is just a tool, like a hammer. If a hammer's too heavy to pick up, then you need to buy a lighter one. Similarly, if the multiplication table is too big to work with, you need a smaller one. Besides, I've removed only the numbers you don't need. For example, the condensed table doesn't include rows or columns for 0, 1, or 2. Here's why:

- Any number multiplied by 0 is 0 (people call this trait the *zero property of multiplication*).

- Any number multiplied by 1 is that number itself (which is why mathematicians call 1 the *multiplicative identity* — because when you multiply any number by 1, the answer is identical to the number you started with).
- Multiplying by 2 is fairly easy; if you can count by 2s — 2, 4, 6, 8, 10, and so forth — you can multiply by 2.

The rest of the numbers I've gotten rid of are redundant. (And not just redundant, but also repeated, extraneous, and unnecessary!) For example, any way you slice it, 3 × 5 and 5 × 3 are both 15 (you can switch the order of the factors because multiplication is commutative — see Chapter 4 for details). In my condensed table, I've simply removed the clutter.

So what's left? Just the numbers you need. These numbers include a gray row and a gray diagonal. The gray row is the 5 times table, which you probably know pretty well. (In fact, the 5s may evoke a childhood memory of running to find a hiding place on a warm spring day while one of your friends counted in a loud voice: 5, 10, 15, 20, ...)

The numbers on the gray diagonal are the square numbers. As I discuss in Chapter 1, when you multiply any number by itself, the result is a square number. You probably know these numbers better than you think.

Getting to know the short multiplication table

In about an hour, you can make huge strides in memorizing the multiplication table. To start, make a set of flash cards that give a multiplication problem on the front and the answer on the back. They may look like Figure 3-1.

Figure 3-1: Both sides of a flash card, with 7 × 6 on the front and 42 on the back.

Illustration by Wiley, Composition Services Graphics

To the nines: A slick trick

Here's a trick to help you remember the 9 times table. To multiply any one-digit number by 9,

1. **Subtract 1 from the number being multiplied by 9 and jot down the answer.**

   For example, suppose you want to multiply 7×9. Here, $7 - 1 = \mathbf{6}$.

2. **Jot down a second number so that, together, the two numbers you wrote add up to 9. You've just written the answer you were looking for.**

   Adding, you get $6 + 3 = \mathbf{9}$. So $7 \times 9 = 63$.

As another example, suppose you want to multiply 8×9:

$8 - 1 = \mathbf{7}$

$7 + 2 = \mathbf{9}$

So $8 \times 9 = 72$.

This trick works for every one-digit number except 0 (but you already know that $0 \times 9 = 0$).

Remember, you need to make only 28 flash cards — one for every example in Table 3-2. Split these 28 into two piles — a "gray" pile with 11 cards and a "white" pile with 17. (You don't have to color the cards gray and white; just keep track of which pile is which, according to the shading in Table 3-2.) Then begin:

1. **5 minutes:** Work with the gray pile, going through it one card at a time. If you get the answer right, put that card on the bottom of the pile. If you get it wrong, put it in the middle so you get another chance at it more quickly.
2. **10 minutes:** Switch to the white pile and work with it in the same way.
3. **15 minutes:** Repeat Steps 1 and 2.

Now take a break. Really — the break is important to rest your brain. Come back later in the day and do the same thing.

When you're done with this exercise, you should find going through all 28 cards with almost no mistakes to be fairly easy. At this point, feel free to make cards for the rest of the standard times table — you know, the cards with all the 0, 1, and 2 times tables on them and the redundant problems — mix all 100 cards together, and amaze your family and friends.

Double digits: Multiplying larger numbers

The main reason to know the multiplication table is so you can more easily multiply larger numbers. For example, suppose you want to multiply 53 × 7. Start by stacking these numbers on top of one another with a line underneath, and then multiply 3 by 7. Because 3 × 7 = 21, write down the 1 and carry the 2:

$$\begin{array}{r} \mathbf{2} \\ \mathbf{53} \\ \underline{\mathbf{\times 7}} \\ \mathbf{1} \end{array}$$

Next, multiply 7 by 5. This time, 5 × 7 = 35. But you also need to add the 2 that you carried over, which makes the result 37. Because 5 and 7 are the last numbers to multiply, you don't have to carry, so write down the 37 — you find that 53 × 7 = 371:

$$\begin{array}{r} 2 \\ \mathbf{53} \\ \underline{\times\ \mathbf{7}} \\ \mathbf{37}1 \end{array}$$

When multiplying larger numbers, the idea is similar. For example, suppose you want to multiply 53 by 47. (The first few steps — multiplying by the 7 in 47 — are the same, so I pick up with the next step.) Now you're ready to multiply by the 4 in 47. But remember that this 4 is in the tens column, so it really means 40. So to begin, put a 0 directly under the 1 in 371:

$$\begin{array}{r} 53 \\ \underline{\times\mathbf{47}} \\ 371 \\ \mathbf{2}0 \end{array}$$

This 0 acts as a placeholder so that this row is arranged properly. (See Chapter 1 for more about placeholding zeros.)

When multiplying by larger numbers with two digits or more, use one placeholding zero when multiplying by the tens digit, two placeholding zeros when multiplying the hundreds digit, three zeros when multiplying by the thousands digit, and so forth.

Now you multiply 3 × 4 to get 12, so write down the 2 and carry the 1:

$$\begin{array}{r} \mathbf{1} \\ \mathbf{5}3 \\ \times \mathbf{4}7 \\ \hline 371 \\ \mathbf{20} \end{array}$$

Continuing, multiply 5 × 4 to get 20, and then add the 1 that you carried over, giving a result of 21:

$$\begin{array}{r} \mathbf{1} \\ \mathbf{5}3 \\ \times \mathbf{4}7 \\ \hline 371 \\ \mathbf{21}20 \end{array}$$

To finish, add the two products (the multiplication results):

$$\begin{array}{r} 53 \\ \times 47 \\ \hline 371 \\ +\ 2120 \\ \hline \mathbf{2,491} \end{array}$$

So 53 × 47 = 2,491.

Doing Division Lickety-Split

The last of the Big Four operations is division. Division literally means splitting things up. For example, suppose you're a parent on a picnic with your three children. You've brought along 12 pretzel sticks as snacks, and want to split them fairly so that each child gets the same number (don't want to cause a fight, right?).

Each child gets four pretzel sticks. This problem tells you that

$$12 \div 3 = 4$$

As with multiplication, division also has more than one sign: the division sign (÷) and the fraction slash (/) or fraction bar (—). So some other ways to write the same information are

$$12/3 = 4 \quad \text{and} \quad \frac{12}{3} = 4$$

Whatever happened to the division table?

Considering how much time teachers spend on the multiplication table, you may wonder why you've never seen a division table. For one thing, the multiplication table focuses on multiplying all the one-digit numbers by each other. This focus doesn't work too well for division because division usually involves at least one number that has more than one digit.

Besides, you can use the multiplication table for division, too, by reversing the way you normally use the table. For example, the multiplication table tells you that $6 \times 7 = 42$. You can reverse this equation to give you these two division problems:

$$42 \div 6 = 7$$

$$42 \div 7 = 6$$

Using the multiplication table in this way takes advantage of the fact that multiplication and division are *inverse operations.* I discuss this important idea further in Chapter 4.

Whichever way you write it, the idea is the same: When you divide 12 pretzel sticks equally among three people, each person gets 4 of them.

When you divide one number by another, the first number is called the *dividend,* the second is called the *divisor,* and the result is the *quotient.* For example, in the division from the earlier example, the dividend is 12, the divisor is 3, and the quotient is 4.

Making short work of long division

In the olden days, knowing how to divide large numbers — for example, 62,997 ÷ 843 — was important. People used *long division,* an organized method for dividing a large number by another number. The process involved dividing, multiplying, subtracting, and dropping numbers down.

But face it — one of the main reasons the pocket calculator was invented was to save 21st-century humans from ever having to do long division again.

Having said that, I need to add that your teacher and math-crazy friends may not agree. Perhaps they just want to make sure you're not completely helpless if your calculator disappears somewhere into your backpack or your desk drawer or the Bermuda Triangle. But if do you get stuck doing page after page of long division against your will, you have my deepest sympathy.

I will go this far, however: Understanding how to do long division with some not-too-horrible numbers is a good idea. In this section, I give you a good start with long division, telling you how to do a division problem that has a one-digit divisor.

Recall that the *divisor* in a division problem is the number that you're dividing by. When you're doing long division, the size of the divisor is your main concern: Small divisors are easy to work with, and large ones are a royal pain.

Suppose you want to find 860 ÷ 5. Start off by writing the problem like this:

$$5\overline{)860}$$

Unlike the other Big Four operations, long division moves from left to right. In this case, you start with the number in the hundreds column (8). To begin, ask how many times 5 goes into 8 — that is, what's 8 ÷ 5? The answer is 1 (with a little bit left over), so write 1 directly above the 8. Now multiply 1 × 5 to get 5, place the answer directly below the 8, and draw a line beneath it:

$$\begin{array}{r@{}l} & \;1 \\ 5 & \overline{)860} \\ & \;\underline{5} \end{array}$$

Subtract 8 – 5 to get 3. (***Note:*** After you subtract, the result should always be smaller than the divisor. If not, you need to write a higher number above the division symbol.) Then bring down the 6 to make the new number 36:

$$\begin{array}{r@{}l} & \;1 \\ 5 & \overline{)860} \\ & \underline{-5} \\ & \;36 \end{array}$$

These steps are one complete cycle — to complete the problem, you just need to repeat them. Now ask how many times 5 goes into 36 — that is, what's 36 ÷ 5? The answer is 7 (with a little left over). Write 7 just above the 6, and then multiply 7 × 5 to get 35; write the answer under 36:

$$\begin{array}{r@{}l} & \;17 \\ 5 & \overline{)860} \\ & \underline{-5} \\ & \;36 \\ & \underline{-35} \end{array}$$

Now subtract to get 36 – 35 = 1; bring down the 0 next to the 1 to make the new number 10:

$$\begin{array}{r} 172 \\ 5\overline{)860} \\ \underline{-5} \\ 36 \\ \underline{-35} \\ 10 \end{array}$$

Another cycle is complete, so begin the next cycle by asking how many times 5 goes into 10 — that is, 10 ÷ 5. The answer this time is 2. Write down the 2 in the answer above the 0. Multiply to get 2 × 5 = 10, and write this answer below the 10:

$$\begin{array}{r} 172 \\ 5\overline{)860} \\ \underline{-5} \\ 36 \\ \underline{-35} \\ 10 \\ \underline{-10} \end{array}$$

Now subtract 10 – 10 = 0. Because you have no more numbers to bring down, you're finished, and here's the answer (that is, the quotient):

$$\begin{array}{r} 172 \\ 5\overline{)860} \\ \underline{-5} \\ 36 \\ \underline{-35} \\ 10 \\ \underline{-10} \\ 0 \end{array}$$

So 860 ÷ 5 = 172.

This problem divides evenly, but many don't. The following section tells you what to do when you run out of numbers to bring down, and Chapter 11 explains how to get a decimal answer.

Getting leftovers: Division with a remainder

Division is different from addition, subtraction, and multiplication in that having a remainder is possible. A *remainder* is simply a portion left over from the division.

The letter *r* indicates that the number that follows is the remainder.

For example, suppose you want to divide seven candy bars between two people without breaking any candy bars into pieces (too messy). So each person receives three candy bars, and one candy bar is left over. This problem shows you the following:

$7 \div 2 = 3$ with a remainder of 1, or 3r1

In long division, the remainder is the number that's left when you no longer have numbers to bring down. The following equation shows that $47 \div 3 =$ 15r2:

$$\begin{array}{r} 15 \\ 3\overline{)47} \\ -3 \\ \hline 17 \\ -15 \\ \hline 2 \end{array}$$

Note that when you're doing division with a small dividend and a large divisor, you always get a quotient of 0 and a remainder of the number you started with:

$$1 \div 2 = 0r1$$
$$14 \div 23 = 0r14$$
$$2{,}000 \div 2{,}001 = 0r2000$$

Part II

Getting a Handle on Whole Numbers

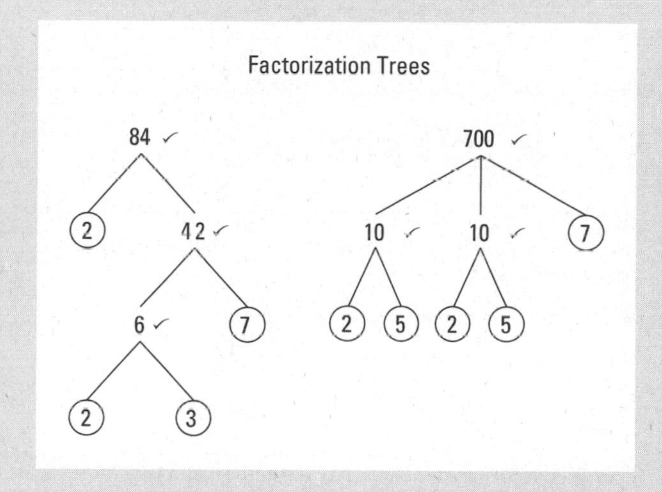

You can use prime factorization to find the greatest common factor (GCF) of difficult numbers. To find out how, go to `www.dummies.com/extras/basicmathandprealgebra`

In this part...

- Add, subtract, multiply, and divide more complex calculations involving negative numbers, inequalities, exponents, square roots, and absolute value
- Build equations and evaluate expressions
- Understand arithmetic word problems
- Employ a few quick tricks to determine whether one number is divisible by another
- Find the factors and multiples of a number, and discover whether a number is prime or composite
- Calculate the greatest common factor (GCF) and the least common multiple (LCM) of a set of numbers

Chapter 4

Putting the Big Four Operations to Work

In This Chapter

- Identifying which operations are inverses of each other
- Knowing the operations that are commutative, associative, and distributive
- Performing the Big Four operations on negative numbers
- Using four symbols for inequality
- Understanding exponents, roots, and absolute values

When you understand the Big Four operations that I cover in Chapter 3 — adding, subtracting, multiplying, and dividing — you can begin to look at math on a whole new level. In this chapter, you extend your understanding of the Big Four operations and move beyond them. I begin by focusing on four important properties of the Big Four operations: inverse operations, commutative operations, associative operations, and distribution. Then I show you how to perform the Big Four on negative numbers.

I continue by introducing you to some important symbols for inequality. Finally, you're ready to move beyond the Big Four by discovering three more advanced operations: exponents (also called *powers*), square roots (also called *radicals*), and absolute values.

Knowing Properties of the Big Four Operations

When you know how to do the Big Four operations — add, subtract, multiply, and divide — you're ready to grasp a few important *properties* of these important operations. Properties are features of the Big Four operations that always apply, no matter what numbers you're working with.

In this section, I introduce you to four important ideas: inverse operations, commutative operations, associative operations, and the distributive property. Understanding these properties can show you hidden connections among the Big Four operations, save you time when calculating, and get you comfortable working with more-abstract concepts in math.

Inverse operations

Each of the Big Four operations has an *inverse* — an operation that undoes it. Addition and subtraction are inverse operations because addition undoes subtraction, and vice versa. For example, here are two equations with inverse operations:

$$1 + 2 = 3$$

$$3 - 2 = 1$$

In the first equation, you start with 1 and add 2 to it, which gives you 3. In the second equation, you have 3 and take away 2 from it, which brings you back to 1. The main idea here is that you're given a starting number — in this case, 1 — and when you add a number and then subtract the same number, you end up again with the starting number. This shows you that subtraction undoes addition.

Similarly, addition undoes subtraction — that is, if you subtract a number and then add the same number, you end up where you started. For example,

$$184 - 10 = 174$$

$$174 + 10 = 184$$

This time, in the first equation, you start with 184 and take away 10 from it, which gives you 174. In the second equation, you have 174 and add 10 to it, which brings you back to 184. In this case, starting with the number 184, when you subtract a number and then add the same number, the addition undoes the subtraction and you end up back at 184.

In the same way, multiplication and division are inverse operations. For example,

$$4 \times 5 = 20$$

$$20 \div 5 = 4$$

This time, you start with the number 4 and multiply it by 5 to get 20. And then you divide 20 by 5 to return to where you started at 4. So division undoes multiplication. Similarly,

$$30 \div 10 = 3$$
$$3 \times 10 = 30$$

Here, you start with 30, divide by 10, and multiply by 10 to end up back at 30. This shows you that multiplication undoes division.

Commutative operations

Addition and multiplication are both commutative operations. *Commutative* means that you can switch around the order of the numbers without changing the result. This property of addition and multiplication is called the *commutative property.* Here's an example of how addition is commutative:

$$3 + 5 = 8 \quad \text{is the same as} \quad 5 + 3 = 8$$

If you start out with 5 books and add 3 books, the result is the same as if you start with 3 books and add 5. In each case, you end up with 8 books.

And here's an example of how multiplication is commutative:

$$2 \times 7 = 14 \quad \text{is the same as} \quad 7 \times 2 = 14$$

If you have 2 children and want to give them each 7 flowers, you need to buy the same number of flowers as someone who has 7 children and wants to give them each 2 flowers. In both cases, someone buys 14 flowers.

In contrast, subtraction and division are *non-commutative* operations. When you switch the order of the numbers, the result changes.

Here's an example of how subtraction is non-commutative:

$$6 - 4 = 2 \quad \text{but} \quad 4 - 6 = -2$$

Subtraction is non-commutative, so if you have $6 and spend $4, the result is *not* the same as if you have $4 and spend $6. In the first case, you still have $2 left over. In the second case, you *owe* $2. In other words, switching the numbers turns the result into a negative number. (I discuss negative numbers later in this chapter.)

And here's an example of how division is non-commutative:

$$5 \div 2 = 2r1 \quad \text{but} \quad 2 \div 5 = 0r2$$

For example, when you have five dog biscuits to divide between two dogs, each dog gets two biscuits and you have one biscuit left over. But when you switch the numbers and try to divide two biscuits among five dogs, you don't have enough biscuits to go around, so each dog gets none and you have two left over.

Associative operations

Addition and multiplication are both *associative operations*, which means that you can group them differently without changing the result. This property of addition and multiplication is also called the *associative property.* Here's an example of how addition is associative. Suppose you want to add 3 + 6 + 2. You can calculate in two ways:

$$\begin{array}{ll} (3+6)+2 & 3+(6+2) \\ =9+2 & =3+8 \\ =11 & =11 \end{array}$$

In the first case, I start by adding 3 + 6 and then add 2. In the second case, I start by adding 6 + 2 and then add 3. Either way, the sum is 11.

And here's an example of how multiplication is associative. Suppose you want to multiply $5 \times 2 \times 4$. You can calculate in two ways:

$$\begin{array}{ll} (5\times2)\times4 & 5\times(2\times4) \\ =10\times4 & =5\times8 \\ =40 & =40 \end{array}$$

In the first case, I start by multiplying 5 × 2 and then multiply by 4. In the second case, I start by multiplying 2 × 4 and then multiply by 5. Either way, the product is 40.

In contrast, subtraction and division are non-associative operations. This means that grouping them in different ways changes the result.

Don't confuse the commutative property with the associative property. The commutative property tells you that it's okay *to switch* two numbers that you're adding or multiplying. The associative property tells you that it's okay to *regroup* three (or more) numbers using parentheses.

Taken together, the commutative and associative properties allow you to completely rearrange and regroup a string of numbers that you're adding or multiplying without changing the result. You'll find the freedom to rearrange expressions as you like to be very useful as you move on to algebra in Part V.

Distribution to lighten the load

If you've ever tried to carry a heavy bag of groceries, you may have found that distributing the contents into two smaller bags is helpful. This same concept also works for multiplication.

In math, *distribution* (also called the *distributive property of multiplication over addition*) allows you to split a large multiplication problem into two smaller ones and add the results to get the answer.

For example, suppose you want to multiply these two numbers:

$$17 \times 101$$

You can go ahead and just multiply them, but distribution provides a different way to think about the problem that you may find easier. Because 101 = 100 + 1, you can split this problem into two easier problems, as follows:

$$= 17 \times (100 + 1)$$
$$= (17 \times 100) + (17 \times 1)$$

You take the number outside the parentheses, multiply it by each number inside the parentheses one at a time, and then add the products. At this point, you may be able to calculate the two multiplications in your head and then add them up easily:

$$= 1{,}700 + 17 = 1{,}717$$

Distribution becomes even more useful when you get to algebra in Part V.

Doing Big Four Operations with Negative Numbers

In Chapter 1, I show you how to use the number line to understand how negative numbers work. In this section, I give you a closer look at how to perform the Big Four operations with negative numbers. Negative numbers result when you subtract a larger number from a smaller one. For example,

$$5 - 8 = -3$$

In real-world applications, negative numbers represent debt. For example, if you have only five chairs to sell but a customer pays for eight of them, you owe her three more chairs. Even though you may have trouble picturing –3 chairs, you still need to account for this debt, and negative numbers are the right tool for the job.

Addition and subtraction with negative numbers

The great secret to adding and subtracting negative numbers is to turn every problem into a series of ups and downs on the number line. When you know how to do this, you find that all these problems are quite simple.

So in this section, I explain how to add and subtract negative numbers on the number line. Don't worry about memorizing every little bit of this procedure. Instead, just follow along so you get a sense of how negative numbers fit onto the number line. (If you need a quick refresher on how the number line works, see Chapter 1.)

Starting with a negative number

When you're adding and subtracting on the number line, starting with a negative number isn't much different from starting with a positive number. For example, suppose you want to calculate –3 + 4. Using the up and down rules, you start at –3 and go up 4:

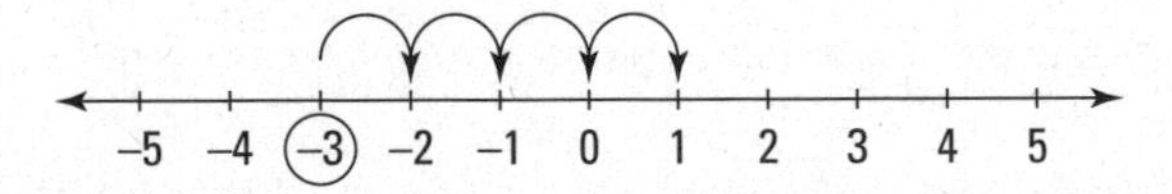

So $-3 + 4 = 1$.

Similarly, suppose you want to calculate –2 – 5. Again, the up and down rules help you out. You're subtracting, so move to the left: start at –2, down 5:

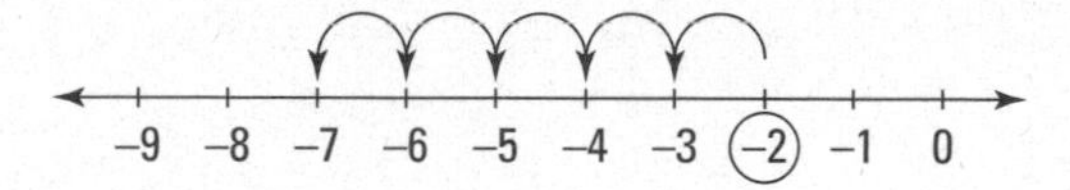

So –2 – 5 = –7.

Adding a negative number

Suppose you want to calculate –2 + –4. You already know to start at –2, but where do you go from there? Here's the up and down rule for adding a negative number:

Adding a negative number is the same as subtracting a positive number — go *down* on the number line.

By this rule, –2 + –4 is the same as –2 – 4, so start at –2, down 4:

–8 –7 –6 –5 –4 –3 –2 –1 0 1 2

So –2 + (–4) = –6.

Note: The problem –2 + –4 can also be written as –2 + (–4). Some people prefer to use this convention so that two operation symbols (– and +) aren't side by side. Don't let it trip you up. The problem is the same.

If you rewrite a subtraction problem as an addition problem — for instance, rewriting 3 – 7 as 3 + (–7) — you can use the commutative and associative properties of addition, which I discuss earlier in this chapter. Just remember to keep the negative sign attached to the number when you rearrange: (–7) + 3.

Subtracting a negative number

The last rule you need to know is how to subtract a negative number. For example, suppose you want to calculate 2 – (–3). Here's the up and down rule:

Subtracting a negative number is the same as adding a positive number — go *up* on the number line.

This rule tells you that 2 – (–3) is the same as 2 + 3, so start at 2, up 3:

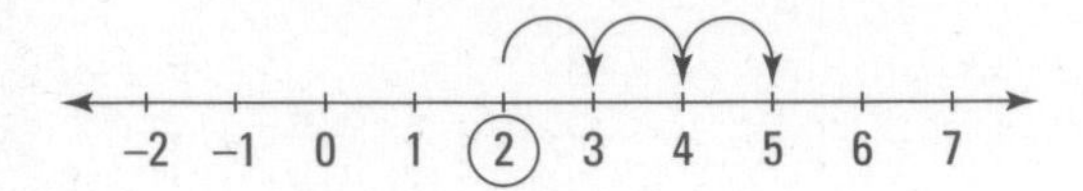

So 2 – (–3) = 5.

When subtracting negative numbers, you can think of the two minus signs canceling each other out to create a positive.

Multiplication and division with negative numbers

Multiplication and division with negative numbers is virtually the same as with positive numbers. The presence of one or more minus signs (–) doesn't change the numerical part of the answer. The only question is whether the sign is positive or negative:

Just remember that when you multiply or divide two numbers,

- If the numbers have the *same sign,* the result is always positive.
- If the numbers have *opposite signs,* the result is always negative.

For example,

$$2 \times 3 = 6 \quad 2 \times -3 = -6$$
$$-2 \times -3 = 6 \quad -2 \times 3 = -6$$

As you can see, the numerical portion of the answer is always 6. The only question is whether the complete answer is 6 or –6. That's where the rule of same or opposite signs comes in.

Another way of thinking of this rule is that the two negatives cancel each other out to make a positive.

Similarly, look at these four division equations:

$$10 \div 2 = 5 \quad 10 \div -2 = -5$$
$$-10 \div -2 = 5 \quad -10 \div 2 = -5$$

In this case, the numerical portion of the answer is always 5. When the signs are the same, the result is positive, and when the signs are different, the result is negative.

Understanding Units

Anything that can be counted is a *unit*. That category is a pretty large one because almost anything that you can name can be counted. You discover more about units of measurement in Chapter 15. For now, just understand that all units can be counted, which means that you can apply the Big Four operations to units.

Adding and subtracting units

Adding and subtracting units isn't very different from adding and subtracting numbers. Just remember that you can add or subtract only when the units are the same. For example,

3 chairs + 2 chairs = 5 chairs

4 oranges – 1 orange = 3 oranges

What happens when you try to add or subtract different units? Here's an example:

3 chairs + 2 tables = ?

The only way you can complete this addition is to make the units the same:

3 pieces of furniture + 2 pieces of furniture = 5 pieces of furniture

Multiplying and dividing units

You can always multiply and divide units by a *number*. For example, suppose you have four chairs and but find that you need twice as many for a party. Here's how you represent this idea in math:

4 chairs × 2 = 8 chairs

Similarly, suppose you have 20 cherries and want to split them among four people. Here's how you represent this idea:

20 cherries ÷ 4 = 5 cherries

But you have to be careful when multiplying or dividing units by units. For example:

2 apples × 3 apples = ? WRONG!

12 hats ÷ 6 hats = ? WRONG!

Neither of these equations makes any sense. In these cases, multiplying or dividing by units is meaningless.

In many cases, however, multiplying and dividing units is okay. For example, multiplying *units of length* (such as inches, miles, or meters) results in *square units*. For example,

$$\begin{aligned} 3 \text{ inches} \times 3 \text{ inches} &= 9 \text{ square inches} \\ 10 \text{ miles} \times 5 \text{ miles} &= 50 \text{ square miles} \\ 100 \text{ meters} \times 200 \text{ meters} &= 20{,}000 \text{ square meters} \end{aligned}$$

You find out more about units of length in Chapter 15. Similarly, here are some examples of when dividing units makes sense:

$$\begin{aligned} 12 \text{ slices of pizza} \div 4 \text{ people} &= 3 \text{ slices of pizza/person} \\ 140 \text{ miles} \div 2 \text{ hours} &= 70 \text{ miles/hour} \end{aligned}$$

In these cases, you read the fraction slash (/) as *per:* slices of pizza *per* person or miles *per* hour. You find out more about multiplying and dividing by units in Chapter 15, when I show you how to convert from one unit of measurement to another.

Understanding Inequalities

Sometimes you want to talk about when two quantities are different. These statements are called *inequalities.* In this section, I discuss six types of inequalities: ≠ (doesn't equal), < (less than), > (greater than), ≤ (less than or equal to), ≥ (greater than or equal to), and ≈ (approximately equals).

Doesn't equal (≠)

The simplest inequality is ≠, which you use when two quantities are not equal. For example,

$$\begin{aligned} 2+2 &\neq 5 \\ 3\times 4 &\neq 34 \\ 999{,}999 &\neq 1{,}000{,}000 \end{aligned}$$

You can read ≠ as "doesn't equal" or "is not equal to." Therefore, read $2+2\neq 5$ as "two plus two doesn't equal five."

Less than ($<$) and greater than ($>$)

The symbol $<$ means *less than.* For example, the following statements are true:

$$4 < 5$$

$$100 < 1{,}000$$

$$2 + 2 < 5$$

Similarly, the symbol $>$ means *greater than.* For example,

$$5 > 4$$

$$100 > 99$$

$$2 + 2 > 3$$

The two symbols $<$ and $>$ are similar and easily confused. Here are two simple ways to remember which is which:

- Notice that the $<$ looks sort of like an *L.* This *L* should remind you that it means *less than.*
- Remember that, in any true statement, the *large* open mouth of the symbol is on the side of the *greater* amount, and the *small* point is on the side of the *lesser* amount.

Less than or equal to ($\leq$) and greater than or equal to ($\geq$)

The symbol $\leq$ means *less than or equal to.* For example, the following statements are true:

$$100 \leq 1{,}000$$

$$2 + 2 \leq 5$$

$$2 + 2 \leq 4$$

Similarly, the symbol $\geq$ means *greater than or equal to.* For example,

$$100 \geq 99$$

$$2 + 2 \geq 3$$

$$2 + 2 \geq 4$$

The symbols $\leq$ and $\geq$ are called *inclusive inequalities* because they *include* (allow) the possibility that both sides are equal. In contrast, the symbols $<$ and $>$ are called *exclusive inequalities* because they *exclude* (don't allow) this possibility.

Approximately equals (≈)

In Chapter 2, I show you how rounding numbers makes large numbers easier to work with. In that chapter, I also introduce ≈, which means *approximately equals.*

For example,

$$49 \approx 50$$
$$1{,}024 \approx 1{,}000$$
$$999{,}999 \approx 1{,}000{,}000$$

You can also use ≈ when you estimate the answer to a problem:

$$1{,}000{,}487 + 2{,}001{,}932 + 5{,}000{,}032$$
$$\approx 1{,}000{,}000 + 2{,}000{,}000 + 5{,}000{,}000$$
$$= 8{,}000{,}000$$

Moving Beyond the Big Four: Exponents, Square Roots, and Absolute Value

In this section, I introduce you to three new operations that you need as you move on with math: exponents, square roots, and absolute value. As with the Big Four operations, these three operations tweak numbers in various ways.

To tell the truth, these three operations have fewer everyday applications than the Big Four. But you'll be seeing a lot more of them as you progress in your study of math. Fortunately, they aren't difficult, so this is a good time to become familiar with them.

Understanding exponents

Exponents (also called *powers*) are shorthand for repeated multiplication. For example, 2^3 means to multiply 2 by itself three times. To do that, use the following notation:

$$2^3 = 2 \times 2 \times 2 = 8$$

In this example, 2 is the *base number* and 3 is the *exponent*. You can read 2^3 as "2 to the third power" or "2 to the power of 3" (or even "2 cubed," which has to do with the formula for finding the value of a cube — see Chapter 16 for details).

Here's another example:

10^5 means to multiply 10 by itself five times

That works out like this:

$$10^5 = 10 \times 10 \times 10 \times 10 \times 10 = 100,000$$

This time, 10 is the base number and 5 is the exponent. Read 10^5 as "10 to the fifth power" or "10 to the power of 5."

When the base number is 10, figuring out any exponent is easy. Just write down a 1 and that many 0s after it:

1 with two 0s	*1 with seven 0s*	*1 with twenty 0s*
$10^2 = 100$	$10^7 = 10,000,000$	$10^{20} = 100,000,000,000,000,000,000$

Exponents with a base number of 10 are important in scientific notation, which I cover in Chapter 14.

The most common exponent is the number 2. When you take any whole number to the power of 2, the result is a square number. (For more information on square numbers, see Chapter 1.) For this reason, taking a number to the power of 2 is called *squaring* that number. You can read 3^2 as "three squared," 4^2 as "four squared," and so forth. Here are some squared numbers:

$$3^2 = 3 \times 3 = 9$$
$$4^2 = 4 \times 4 = 16$$
$$5^2 = 5 \times 5 = 25$$

Any number (except 0) raised to the 0 power equals 1. So 1^0, 37^0, and $999,999^0$ are equivalent, or equal, because they all equal 1.

Discovering your roots

Earlier in this chapter, in "Knowing Properties of the Big Four Operations," I show you how addition and subtraction are inverse operations. I also show you how multiplication and division are inverse operations. In a similar way, roots are the inverse operation of exponents.

The most common root is the square root. A *square root* undoes an exponent of 2. For example,

$$3^2 = 3 \times 3 = 9, \text{ so } \sqrt{9} = 3$$
$$4^2 = 4 \times 4 = 16, \text{ so } \sqrt{16} = 4$$
$$5^2 = 5 \times 5 = 25, \text{ so } \sqrt{25} = 5$$

You can read the symbol $\sqrt{\ }$ either as "the square root of" or as "radical." So read $\sqrt{9}$ as either "the square root of 9" or "radical 9."

As you can see, when you take the square root of any square number, the result is the number that you multiplied by itself to get that square number in the first place. For example, to find $\sqrt{100}$, you ask the question, "What number when multiplied by itself equals 100?" The answer here is 10 because

$$10^2 = 10 \times 10 = 100, \text{ so } \sqrt{100} = 10$$

You probably won't use square roots much until you get to algebra, but at that point, they become handy.

Figuring out absolute value

The *absolute value* of a number is the positive value of that number. It tells you how far away from 0 a number is on the number line. The symbol for absolute value is a set of vertical bars.

Taking the absolute value of a positive number doesn't change that number's value. For example,

$$|3| = 3$$
$$|12| = 12$$
$$|145| = 145$$

However, taking the absolute value of a negative number changes it to a positive number:

$$|-5| = 5$$

$$|-10| = 10$$

$$|-212| = 212$$

Finally, the absolute value of 0 is simply 0:

$$|0| = 0$$

Chapter 5

A Question of Values: Evaluating Arithmetic Expressions

In This Chapter

- Understanding the Three E's of math — equations, expressions, and evaluation
- Using order of precedence to evaluate expressions containing the Big Four operations
- Working with expressions that contain exponents
- Evaluating expressions with parentheses

In this chapter, I introduce you to what I call the Three E's of math: equations, expressions, and evaluation. You'll likely find the Three E's of math familiar because, whether you realize it or not, you've been using them for a long time. Whenever you add up the cost of several items at the store, balance your checkbook, or figure out the area of your room, you're evaluating expressions and setting up equations. In this section, I shed light on this stuff and give you a new way to look at it.

You probably already know that an *equation* is a mathematical statement that has an equals sign (=) — for example, $1 + 1 = 2$. An *expression* is a string of mathematical symbols that can be placed on one side of an equation — for example, $1 + 1$. And *evaluation* is finding out the *value* of an expression as a number — for example, finding out that the expression $1 + 1$ is equal to the number 2.

Throughout the rest of the chapter, I show you how to turn expressions into numbers using a set of rules called the *order of operations* (or *order of precedence*). These rules look complicated, but I break them down so you can see for yourself what to do next in any situation.

Seeking Equality for All: Equations

An *equation* is a mathematical statement that tells you that two things have the same value — in other words, it's a statement with an equals sign. The equation is one of the most important concepts in mathematics because it allows you to boil down a bunch of complicated information into a single number.

Mathematical equations come in a lot of varieties: arithmetic equations, algebraic equations, differential equations, partial differential equations, Diophantine equations, and many more. In this book, I look at only two types: arithmetic equations and algebraic equations.

In this chapter, I discuss only *arithmetic equations,* which are equations involving numbers, the Big Four operations, and the other basic operations I introduce in Chapter 4 (absolute values, exponents, and roots). In Part V, I introduce you to algebraic equations. Here are a few examples of simple arithmetic equations:

$$2+2=4$$
$$3\times 4=12$$
$$20\div 2=10$$

And here are a few examples of more-complicated arithmetic equations:

$$1{,}000-1-1-1=997$$
$$(3+5)\div(9-7)=4$$
$$4^2-\sqrt{256}=(791-842)\times 0$$

Three properties of equality

Three properties of equality are *reflexivity, symmetry,* and *transitivity:*

- **Reflexivity** says that everything is equal to itself. For example,

  $1=1 \quad 23=23 \quad 1{,}000{,}007=1{,}000{,}007$

- **Symmetry** says that you can switch the order in which things are equal. For example,

  $4\times 5=20$, so $20=4\times 5$

- **Transitivity** says that if something is equal to two other things, then those two other things are equal to each other. For example,

  $3+1=4$ and $4=2\times 2$, so $3+1=2\times 2$

Because equality has all three of these properties, mathematicians call equality an *equivalence* relation. The inequalities that I introduce in Chapter 4 ($\neq$, $>$, $<$, and $\approx$) don't necessarily share all these properties.

Hey, it's just an expression

An *expression* is any string of mathematical symbols that can be placed on one side of an equation. Mathematical expressions, just like equations, come in a lot of varieties. In this chapter, I focus only on *arithmetic expressions,* which are expressions that contain numbers, the Big Four operations, and a few other basic operations (see Chapter 4). In Part V, I introduce you to algebraic expressions. Here are a few examples of simple expressions:

$$2+2$$
$$-17+(-1)$$
$$14\div 7$$

And here are a few examples of more-complicated expressions:

$$(88-23)\div 13$$
$$100+2-3\times 17$$
$$\sqrt{441}+|-2^3|$$

Evaluating the situation

At the root of the word *evaluation* is the word *value.* In other words, when you evaluate something, you find its value. Evaluating an expression is also referred to as *simplifying, solving,* or *finding the value of an expression.* The words may change, but the idea is the same — boiling down a string of numbers and math symbols to a single number.

When you evaluate an arithmetic expression, you simplify it to a single numerical value — in other words, you find the number that it's equal to. For example, evaluate the following arithmetic expression:

$$7\times 5$$

How? Simplify it to a single number:

$$35$$

Putting the Three E's together

I'm sure you're dying to know how the Three E's — equations, expressions, and evaluation — are all connected. *Evaluation* allows you to take an *expression* containing more than one number and reduce it to a single number. Then you can make an *equation,* using an equals sign, to connect the expression and the number. For example, here's an *expression* containing four numbers:

$$1 + 2 + 3 + 4$$

When you *evaluate* it, you reduce it to a single number:

$$10$$

And now you can make an *equation* by connecting the expression and the number with an equals sign:

$$1 + 2 + 3 + 4 = 10$$

Introducing Order of Operations

When you were a kid, did you ever try putting on your shoes first and then your socks? If you did, you probably discovered this simple rule:

1. **Put on socks.**
2. **Put on shoes.**

Thus, you have an order of operations: The socks have to go on your feet before your shoes. So in the act of putting on your shoes and socks, your socks have precedence over your shoes. A simple rule to follow, right?

In this section, I outline a similar set of rules for evaluating expressions, called the *order of operations* (sometimes called *order of precedence*). Don't let the long name throw you. Order of operations is just a set of rules to make sure you get your socks and shoes on in the right order, mathematically speaking, so you always get the right answer.

Note: Through most of this book, I introduce overarching themes at the beginning of each section and then explain them later in the chapter instead of building them and finally revealing the result. But order of operations is a bit too confusing to present that way. Instead, I start with a list of four rules and go into more detail about them later in the chapter. Don't let the complexity of these rules scare you off before you work through them!

Evaluate arithmetic expressions from left to right according to the following order of operations:

1. **Parentheses**
2. **Exponents**

3. **Multiplication and division**
4. **Addition and subtraction**

Don't worry about memorizing this list right now. I break it to you slowly in the remaining sections of this chapter, starting from the bottom and working toward the top, as follows:

- In "Applying order of operations to Big Four expressions," I show Steps 3 and 4 — how to evaluate expressions with any combination of addition, subtraction, multiplication, and division.
- In "Using order of operations in expressions with exponents," I show you how Step 2 fits in — how to evaluate expressions with Big Four operations plus exponents, square roots, and absolute value.
- In "Understanding order of operations in expressions with parentheses," I show you how Step 1 fits in — how to evaluate all the expressions I explain plus expressions with parentheses.

Applying order of operations to Big Four expressions

As I explain earlier in this chapter, evaluating an expression is just simplifying it to a single number. Now I get you started on the basics of evaluating expressions that contain any combination of the Big Four operations — adding, subtracting, multiplying, and dividing. (For more on the Big Four, see Chapter 3.) Generally speaking, the Big Four expressions come in the three types in Table 5-1.

Table 5-1 The Three Types of Big Four Expressions

Expression	*Example*	*Rule*
Contains only addition and subtraction	$12 + 7 - 6 - 3 + 8$	Evaluate left to right.
Contains only multiplication and division	$18 \div 3 \times 7 \div 14$	Evaluate left to right.
Mixed-operator expression: contains a combination of addition/subtraction and multiplication/division	$9 + 6 \div 3$	1. Evaluate multiplication and division left to right. 2. Evaluate addition and subtraction left to right.

In this section, I show you how to identify and evaluate all three types of expressions.

Expressions with only addition and subtraction

Some expressions contain only addition and subtraction. When this is the case, the rule for evaluating the expression is simple.

When an expression contains only addition and subtraction, evaluate it step by step from left to right. For example, suppose you want to evaluate this expression:

$$17 - 5 + 3 - 8$$

Because the only operations are addition and subtraction, you can evaluate from left to right, starting with 17 – 5:

$$= 12 + 3 - 8$$

As you can see, the number 12 replaces 17 – 5. Now the expression has three numbers instead of four. Next, evaluate 12 + 3:

$$= 15 - 8$$

This step breaks down the expression to two numbers, which you can evaluate easily:

$$= 7$$

So $17 - 5 + 3 - 8 = 7$.

Expressions with only multiplication and division

Some expressions contain only multiplication and division. When this is the case, the rule for evaluating the expression is pretty straightforward.

When an expression contains only multiplication and division, evaluate it step by step from left to right. Suppose you want to evaluate this expression:

$$9 \times 2 \div 6 \div 3 \times 2$$

Again, the expression contains only multiplication and division, so you can move from left to right, starting with 9×2:

$$= 18 \div 6 \div 3 \times 2$$
$$= 3 \div 3 \times 2$$
$$= 1 \times 2$$
$$= 2$$

Notice that the expression shrinks one number at a time until all that's left is 2. So $9\times2\div6\div3\times2=2$.

Here's another quick example:

$$-2\times6\div-4$$

Even though this expression has some negative numbers, the only operations it contains are multiplication and division. So you can evaluate it in two steps from left to right (remembering the rules for multiplying and dividing with negative numbers that I show you in Chapter 4):

$$=-2\times6\div-4$$
$$=-12\div-4$$
$$=3$$

Thus, $-2\times6\div-4=3$.

Mixed-operator expressions

Often an expression contains

- At least one addition or subtraction operator
- At least one multiplication or division operator

I call these *mixed-operator expressions.* To evaluate them, you need some stronger medicine.

Evaluate mixed-operator expressions as follows:

1. **Evaluate the multiplication and division from left to right.**
2. **Evaluate the addition and subtraction from left to right.**

For example, suppose you want to evaluate the following expression:

$$5+3\times2+8\div4$$

As you can see, this expression contains addition, multiplication, and division, so it's a mixed-operator expression. To evaluate it, start by underlining the multiplication and division in the expression:

$$5+\underline{3\times2}+\underline{8\div4}$$

Now evaluate what you've underlined from left to right:

$$= 5 + 6 + \underline{8 \div 4}$$
$$= 5 + 6 + 2$$

At this point, you're left with an expression that contains only addition, so you can evaluate it from left to right:

$$= 11 + 2$$
$$= 13$$

Thus, $5 + 3 \times 2 + 8 \div 4 = 13$.

Using order of operations in expressions with exponents

Here's what you need to know to evaluate expressions that have exponents (see Chapter 4 for info on exponents).

Evaluate exponents from left to right *before* you begin evaluating Big Four operations (adding, subtracting, multiplying, and dividing).

The trick here is to turn the expression into a Big Four expression and then use what I show you earlier in "Applying order of operations to Big Four expressions." For example, suppose you want to evaluate the following:

$$3 + 5^2 - 6$$

First, evaluate the exponent:

$$3 + 25 - 6$$

At this point, the expression contains only addition and subtraction, so you can evaluate it from left to right in two steps:

$$= 28 - 6$$
$$= 22$$

So $3 + 5^2 - 6 = 22$.

Understanding order of precedence in expressions with parentheses

In math, parentheses — () — are often used to group together parts of an expression. When it comes to evaluating expressions, here's what you need to know about parentheses.

To evaluate expressions that contain parentheses,

1. **Evaluate the contents of parentheses from the inside out.**
2. **Evaluate the rest of the expression.**

Big Four expressions with parentheses

Similarly, suppose you want to evaluate $(1+15\div5)+(3-6)\times5$. This expression contains two sets of parentheses, so evaluate these from left to right. Notice that the first set of parentheses contains a mixed-operator expression, so evaluate this in two steps, starting with the division:

$$=(1+3)+(3-6)\times5$$
$$=4+(3-6)\times5$$

Now evaluate the contents of the second set of parentheses:

$$=4+-3\times5$$

Now you have a mixed-operator expression, so evaluate the multiplication (-3×5) first:

$$=4+-15$$

Finally, evaluate the addition:

$$=-11$$

So $(1+15\div5)+(3-6)\times5=-11$.

Expressions with exponents and parentheses

As another example, try this out:

$$1+(3-6^2\div9)\times2^2$$

Start by working with *only* what's inside the parentheses. The first part to evaluate there is the exponent, 6^2:

$$= 1 + (3 - 36 \div 9) \times 2^2$$

Continue working inside the parentheses by evaluating the division $36 \div 9$:

$$= 1 + (3 - 4) \times 2^2$$

Now you can get rid of the parentheses altogether:

$$= 1 - 1 \times 2^2$$

At this point, what's left is an expression with an exponent. This expression takes three steps, starting with the exponent:

$$= 1 - 1 \times 4$$
$$= 1 - 4$$
$$= -3$$

So $1 + (3 - 6^2 \div 9) \times 2^2 = -3$.

Expressions with parentheses raised to an exponent

Sometimes the entire contents of a set of parentheses are raised to an exponent. In this case, evaluate the contents of the parentheses *before* evaluating the exponent, as usual. Here's an example:

$$(7 - 5)^3$$

First, evaluate 7 – 5:

$$= 2^3$$

With the parentheses removed, you're ready to evaluate the exponent:

$$= 8$$

Once in a rare while, the exponent itself contains parentheses. As always, evaluate what's in the parentheses first. For example,

$$21^{(19 + \underline{3 \times -6})}$$

This time, the smaller expression inside the parentheses is a mixed-operator expression. I've underlined the part that you need to evaluate first:

$$= 21^{(19-18)}$$

Now you can finish off what's inside the parentheses:

$$= 21^1$$

At this point, all that's left is a very simple exponent:

$$= 21$$

So $21^{(19+3\times -6)} = 21$.

Note: Technically, you don't need to put parentheses around the exponent. If you see an expression in the exponent, treat it as though it has parentheses around it. In other words, $21^{19+3\times -6} = 21$ means the same as $21^{(19+3\times -6)} = 21$.

Expressions with nested parentheses

Occasionally, an expression has *nested parentheses,* or one or more sets of parentheses inside another set. Here I give you the rule for handling nested parentheses.

When evaluating an expression with nested parentheses, evaluate what's inside the *innermost* set of parentheses first and work your way toward the *outermost* parentheses.

For example, suppose you want to evaluate the following expression:

$$2 + (9 - (\underline{7 - 3}))$$

I underlined the contents of the innermost set of parentheses, so evaluate these contents first:

$$= 2 + (9 - 4)$$

Next, evaluate what's inside the remaining set of parentheses:

$$= 2 + 5$$

Now you can finish things off easily:

$$= 7$$

So $2 + (9 - (7 - 3)) = 7$.

As a final example, here's an expression that requires everything from this chapter:

$$4+(-7\times(2^{\underline{(5-1)}}-4\times6))$$

This expression is about as complicated as you're ever likely to see in pre-algebra: one set of parentheses containing another set, which contains a third set. To start you off, I underlined what's deep inside this third set of parentheses. This is where you begin evaluating:

$$=4+(-7\times(\underline{2^4-4\times6}))$$

What's left is one set of parentheses inside another set. Again, work from the inside out. The smaller expression here is $2^4-4\times6$, so evaluate the exponent first, then the multiplication, and finally the subtraction:

$$=4+(-7\times(\underline{16-4\times6}))$$
$$=4+(-7\times(\underline{16-24}))$$
$$=4+(-7\times-8)$$

Only one more set of parentheses to go:

$$=4+56$$

At this point, finishing up is easy:

$$=60$$

Therefore, $4+(-7\times(2^{\underline{(5-1)}}-4\times6))=60$.

As I say earlier in this section, this problem is about as hard as they come at this stage of math. Copy it down and try solving it step by step with the book closed.

Chapter 6

Say What? Turning Words into Numbers

In This Chapter

- Dispelling myths about word problems
- Knowing the four steps to solving a word problem
- Jotting down simple word equations that condense the important information
- Writing more-complex word equations
- Plugging numbers into the word equations to solve the problem
- Attacking more-complex word problems with confidence

The very mention of word problems — or story problems, as they're sometimes called — is enough to send a cold shiver of terror into the bones of the average math student. Many would rather swim across a moat full of hungry crocodiles than "figure out how many bushels of corn Farmer Brown picked" or "help Aunt Sylvia decide how many cookies to bake." But word problems help you understand the logic behind setting up equations in real-life situations, making math actually useful — even if the scenarios in the word problems you practice on are pretty far-fetched.

In this chapter, I dispel a few myths about word problems. Then I show you how to solve a word problem in four simple steps. After you understand the basics, I show you how to solve more-complex problems. Some of these problems have longer numbers to calculate, and others may have more complicated stories. In either case, you can see how to work through them step by step.

Dispelling Two Myths about Word Problems

Here are two common myths about word problems:

- Word problems are always hard.
- Word problems are only for school — after that, you don't need them.

Both of these ideas are untrue. But they're so common that I want to address them head-on.

Word problems aren't always hard

Word problems don't have to be hard. For example, here's a word problem that you may have run into in first grade:

> Adam had 4 apples. Then Brenda gave him 5 more apples. How many apples does Adam have now?

You can probably do the math in your head, but when you were starting out in math, you may have written it down:

$$4 + 5 = 9$$

Finally, if you had one of those teachers who made you write out your answer in complete sentences, you wrote "Adam has 9 apples." (Of course, if you were the class clown, you probably wrote, "Adam doesn't have any apples because he ate them all.")

Word problems seem hard when they get too complex to solve in your head and you don't have a system for solving them. In this chapter, I give you a system and show you how to apply it to problems of increasing difficulty. And in Chapters 13, 18, and 23, I give you further practice solving more difficult word problems.

Word problems are useful

In the real world, math rarely comes in the form of equations. It comes in the form of situations that are very similar to word problems.

Whenever you paint a room, prepare a budget, bake a double batch of oatmeal cookies, estimate the cost of a vacation, buy wood to build a shelf, do your taxes, or weigh the pros and cons of buying a car versus leasing one, you need math. And the math skill you need most is understanding how to turn the *situation* you're facing into numbers that you calculate.

Word problems give you practice turning situations — or stories — into numbers.

Solving Basic Word Problems

Generally, solving a word problem involves four steps:

1. **Read through the problem and set up a *word equation* — that is, an equation that contains words as well as numbers.**
2. **Plug in numbers in place of words wherever possible to set up a regular math equation.**
3. **Use math to solve the equation.**
4. **Answer the question the problem asks.**

Most of this book is about Step 3. This chapter and Chapters 13, 18, and 23 are all about Steps 1 and 2. I show you how to break down a word problem sentence by sentence, jot down the information you need to solve the problem, and then substitute numbers for words to set up an equation.

When you know how to turn a word problem into an equation, the hard part is done. Then you can use the rest of what you find in this book to figure out how to do Step 3 — solve the equation. From there, Step 4 is usually pretty easy, though at the end of each example, I make sure you understand how to do it.

Turning word problems into word equations

The first step to solving a word problem is reading it and putting the information you find into a useful form. In this section, I show you how to squeeze the juice out of a word problem and leave the pits behind!

Jotting down information as word equations

Most word problems give you information about numbers, telling you exactly how much, how many, how fast, how big, and so forth. Here are some examples:

> Nunu is spinning 17 plates.
>
> The width of the house is 80 feet.
>
> If the local train is going 25 miles per hour …

You need this information to solve the problem. And paper is cheap, so don't be afraid to use it. (If you're concerned about trees, write on the back of all that junk mail you get.) Have a piece of scrap paper handy and jot down a few notes as you read through a word problem.

For example, here's how you can jot down "Nunu is spinning 17 plates":

> Nunu = 17

Here's how to note that "the width of the house is 80 feet":

> width = 80

The third example tells you, "If the local train is going 25 miles per hour… ." So you can jot down the following:

> local = 25

Don't let the word *if* confuse you. When a problem says "If so-and-so were true …" and then asks you a question, assume that it *is* true and use this information to answer the question.

When you jot down information this way, you're really turning words into a more useful form called a *word equation.* A word equation has an equals sign like a math equation, but it contains both words and numbers.

Writing relationships: Turning more-complex statements into word equations

When you start doing word problems, you notice that certain words and phrases show up over and over again. For example,

> Bobo is spinning five fewer plates than Nunu.
>
> The height of a house is half as long as its width.
>
> The express train is moving three times faster than the local train.

You've probably seen statements such as these in word problems since you were first doing math. Statements like these look like English, but they're really math, so spotting them is important. You can represent each of these types of statements as word equations that also use Big Four operations. Look again at the first example:

Bobo is spinning five fewer plates than Nunu.

You don't know the number of plates that either Bobo or Nunu is spinning. But you know that these two numbers are related.

You can express this relationship like this:

Bobo + 5 = Nunu

This word equation is shorter than the statement it came from. And as you see in the next section, word equations are easy to turn into the math you need to solve the problem.

Here's another example:

The height of a house is half as long as its width.

You don't know the width or height of the house, but you know that these numbers are connected.

You can express this relationship between the width and height of the house as the following word equation:

height = width ÷ 2

With the same type of thinking, you can express "The express train is moving three times faster than the local train" as this word equation:

express = 3 × local

As you can see, each of the examples allows you to set up a word equation using one of the Big Four operations — adding, subtracting, multiplying, and dividing.

Figuring out what the problem's asking

The end of a word problem usually contains the question you need to answer to solve the problem. You can use word equations to clarify this question so you know right from the start what you're looking for.

For example, you can write the question, "All together, how many plates are Bobo and Nunu spinning?" as

Bobo + Nunu = ?

You can write the question "How tall is the house" as:

height = ?

Finally, you can rephrase the question "What's the difference in speed between the express train and the local train?" in this way:

express – local = ?

Plugging in numbers for words

After you've written out a bunch of word equations, you have the facts you need in a form you can use. You can often solve the problem by plugging numbers from one word equation into another. In this section, I show you how to use the word equations you built in the last section to solve three problems.

Example: Send in the clowns

Some problems involve simple addition or subtraction. Here's an example:

> Bobo is spinning five fewer plates than Nunu. (Bobo dropped a few.) Nunu is spinning 17 plates. All together, how many plates are Bobo and Nunu spinning?

Here's what you have already, just from reading the problem:

Nunu = 17

Bobo + 5 = Nunu

Plugging in the information gives you the following:

Bobo + 5 = 17

If you see how many plates Bobo is spinning, feel free to jump ahead. If not, here's how you rewrite the addition equation as a subtraction equation (see Chapter 4 for details):

Bobo = 17 – 5 = 12

The problem wants you to find out how many plates the two clowns are spinning together. So you need to find out the following:

Bobo + Nunu = ?

Just plug in the numbers, substituting 12 for Bobo and 17 for Nunu:

$$12 + 17 = 29$$

So Bobo and Nunu are spinning 29 plates.

Example: Our house in the middle of our street

At times, a problem notes relationships that require you to use multiplication or division. Here's an example:

> The height of a house is half as long as its width, and the width of the house is 80 feet. How tall is the house?

You already have a head start from what you determined earlier:

$$\text{width} = 80$$
$$\text{height} = \text{width} \div 2$$

You can plug in information as follows, substituting 80 for the word *width*:

$$\text{height} = 80 \div 2 = 40$$

So you know that the height of the house is 40 feet.

Example: I hear the train a-comin'

Pay careful attention to what the question is asking. You may have to set up more than one equation. Here's an example:

> The express train is moving three times faster than the local train. If the local train is going 25 miles per hour, what's the difference in speed between the express train and the local train?

Here's what you have so far:

$$\text{local} = 25$$
$$\text{express} = 3 \times \text{local}$$

Plug in the information you need:

$$\text{express} = 3 \times 25 = 75$$

In this problem, the question at the end asks you to find the difference in speed between the express train and the local train. Finding the difference between two numbers is subtraction, so here's what you want to find:

$$\text{express} - \text{local} = ?$$

You can get what you need to know by plugging in the information you've already found:

75 – 25 = 50

Therefore, the difference in speed between the express train and the local train is 50 miles per hour.

Solving More-Complex Word Problems

The skills I show you previously in "Solving Basic Word Problems" are important for solving any word problem because they streamline the process and make it simpler. What's more, you can use those same skills to find your way through more complex problems. Problems become more complex when

- The calculations become harder. (For example, instead of a dress costing $30, it costs $29.95.)
- The amount of information in the problem increases. (For example, instead of two clowns, you have five.)

Don't let problems like these scare you. In this section, I show you how to use your new problem-solving skills to solve more-difficult word problems.

When numbers get serious

A lot of problems that look tough aren't much more difficult than the problems I show you in the previous sections. For example, consider this problem:

> Aunt Effie has $732.84 hidden in her pillowcase, and Aunt Jezebel has $234.19 less than Aunt Effie has. How much money do the two women have all together?

One question you may have is how these women ever get any sleep with all that change clinking around under their heads. But moving on to the math, even though the numbers are larger, the principle is still the same as in problems in the earlier sections. Start reading from the beginning: "Aunt Effie has $732.84 . . . ." This text is just information to jot down as a simple word equation:

Effie = $732.84

Continuing, you read, "Aunt Jezebel has *$234.19 less than* Aunt Effie has." It's another statement you can write as a word equation:

Jezebel = Effie – $234.19

Now you can plug in the number $732.84 where you see Aunt Effie's name in the equation:

Jezebel = $732.84 – $234.19

So far, the big numbers haven't been any trouble. At this point, though, you probably need to stop to do the subtraction:

$$\begin{array}{r} \$732.84 \\ -\ \$234.19 \\ \hline \$498.65 \end{array}$$

Now you can jot this information down, as always:

Jezebel = $498.65

The question at the end of the problem asks you to find out how much money the two women have all together. Here's how to represent this question as an equation:

Effie + Jezebel = ?

You can plug information into this equation:

$732.84 + $498.65 = ?

Again, because the numbers are large, you probably have to stop to do the math:

$$\begin{array}{r} \$732.84 \\ +\ \$498.65 \\ \hline \$1231.49 \end{array}$$

So all together, Aunt Effie and Aunt Jezebel have $1,231.49.

As you can see, the procedure for solving this problem is basically the same as for the simpler problems in the earlier sections. The only difference is that you have to stop to do some addition and subtraction.

Too much information

When the going gets tough, knowing the system for writing word equations really becomes helpful. Here's a word problem that's designed to scare you off — but with your new skills, you're ready for it:

> Four women collected money to save the endangered Salt Creek tiger beetle. Keisha collected $160, Brie collected $50 more than Keisha, Amy

collected twice as much as Brie, and together Amy and Sophia collected $700. How much money did the four women collect all together?

If you try to do this problem all in your head, you'll probably get confused. Instead, take it line by line and just jot down word equations as I discuss earlier in this chapter.

First, "Keisha collected $160." So jot down the following:

Keisha = 160

Next, "Brie collected $50 dollars more than Keisha," so write

Brie = Keisha + 50

After that, "Amy collected twice as much as Brie":

Amy = Brie × 2

Finally, "together, Amy and Sophia collected $700":

Amy + Sophia = 700

That's all the information the problem gives you, so now you can start working with it. Keisha collected $160, so you can plug in 160 anywhere you find Keisha's name:

Brie = 160 + 50 = 210

Now you know how much Brie collected, so you can plug this information into the next equation:

Amy = 210 × 2 = 420

This equation tells you how much Amy collected, so you can plug this number into the last equation:

420 + Sophia = 700

To solve this problem, change it from addition to subtraction using inverse operations, as I show you in Chapter 4:

Sophia = 700 – 420 = 280

Now that you know how much money each woman collected, you can answer the question at the end of the problem:

Keisha + Brie + Amy + Sophia = ?

You can plug in this information easily:

$$160 + 210 + 420 + 280 = 1{,}070$$

So you can conclude that the four women collected $1,070 all together.

Putting it all together

Here's one final example putting together everything from this chapter. Try writing down this problem and working it through step by step on your own. If you get stuck, come back here. When you can solve it from beginning to end with the book closed, you'll have a good grasp of how to solve word problems:

> On a recent shopping trip, Travis bought six shirts for $19.95 each and two pairs of pants for $34.60 each. He then bought a jacket that cost $37.08 less than he paid for both pairs of pants. If he paid the cashier with three $100 bills, how much change did he receive?

On the first read-through, you may wonder how Travis found a store that prices jackets that way. Believe me — it was quite a challenge. Anyway, back to the problem. You can jot down the following word equations:

$$\begin{aligned} \text{shirts} &= \$19.95 \times 6 \\ \text{pants} &= \$34.60 \times 2 \\ \text{jacket} &= \text{pants} - \$37.08 \end{aligned}$$

The numbers in this problem are probably longer than you can solve in your head, so they require some attention:

$$\begin{array}{rr} \$19.95 & \$34.60 \\ \times \quad 6 & \times \quad 2 \\ \hline \$119.70 & \$69.20 \end{array}$$

With this done, you can fill in some more information:

$$\begin{aligned} \text{shirts} &= \$119.70 \\ \text{pants} &= \$69.20 \\ \text{jacket} &= \text{pants} - \$37.08 \end{aligned}$$

Now you can plug in $69.20 for *pants:*

$$\text{jacket} = \$69.20 - \$37.08$$

Again, because the numbers are long, you need to solve this equation separately:

$$\begin{array}{r} \$69.20 \\ -\ \$37.08 \\ \hline \$32.12 \end{array}$$

This equation gives you the price of the jacket:

jacket = $32.12

Now that you have the price of the shirts, pants, and jacket, you can find out how much Travis spent:

amount Travis spent = $119.70 + $69.20 + $32.12

Again, you have another equation to solve:

$$\begin{array}{r} \$119.70 \\ \$69.20 \\ +\ \$32.12 \\ \hline \$221.02 \end{array}$$

So you can jot down the following:

amount Travis spent = $221.02

The problem is asking you to find out how much change Travis received from $300, so jot this down:

change = $300 – amount Travis spent

You can plug in the amount that Travis spent:

change = $300 – $221.02

And do just one more equation:

$$\begin{array}{r} \$300.00 \\ -\ \$221.02 \\ \hline \$78.98 \end{array}$$

So you can jot down the answer:

change = $78.98

Therefore, Travis received $78.98 in change.

Chapter 7

Divisibility

In This Chapter

- Finding out whether a number is divisible by 2, 3, 5, 9, 10, or 11
- Seeing the difference between prime numbers and composite numbers

When one number is *divisible* by another, you can divide the first number by the second number without getting a remainder (see Chapter 3 for details on division). In this chapter, I explore divisibility from a variety of angles.

To start, I show you a bunch of handy tricks for discovering whether one number is divisible by another without actually doing the division. (In fact, you don't find long division anywhere in this chapter!) After that, I talk about prime numbers and composite numbers (which I introduce briefly in Chapter 1).

This discussion, plus what follows in Chapter 8, can help make your encounter with fractions in Part III a lot friendlier.

Knowing the Divisibility Tricks

As you begin to work with fractions in Part III, the question of whether one number is divisible by another comes up a lot. In this section, I give you a bunch of time-saving tricks for finding out whether one number is divisible by another without actually making you do the division.

Counting everyone in: Numbers you can divide everything by

Every number is divisible by 1. As you can see, when you divide any number by 1, the answer is the number itself, with no remainder:

$$2 \div 1 = 2$$
$$17 \div 1 = 17$$
$$431 \div 1 = 431$$

Similarly, every number (except 0) is divisible by itself. Clearly, when you divide any number by itself, the answer is 1:

$$5 \div 5 = 1$$
$$28 \div 28 = 1$$
$$873 \div 873 = 1$$

You can't divide any number by 0. Mathematicians say that dividing by 0 is *undefined.*

In the end: Looking at the final digits

You can tell whether a number is divisible by 2, 5, 10, 100, or 1,000 simply by looking at how the number ends — no calculations required.

Divisible by 2

Every even number — that is, every number that ends in 2, 4, 6, 8, or 0 — is divisible by 2. For example, the following bolded numbers are divisible by 2:

$$\mathbf{6} \div 2 = 3$$
$$\mathbf{22} \div 2 = 11$$
$$\mathbf{538} \div 2 = 269$$
$$\mathbf{6{,}790} \div 2 = 3{,}395$$
$$\mathbf{77{,}144} \div 2 = 38{,}572$$
$$\mathbf{212{,}116} \div 2 = 106{,}058$$

Divisible by 5

Every number that ends in either 5 or 0 is divisible by 5. The following bolded numbers are divisible by 5:

$$\mathbf{15} \div 5 = 3$$
$$\mathbf{625} \div 5 = 125$$
$$\mathbf{6,970} \div 5 = 1,394$$
$$\mathbf{44,440} \div 5 = 8,888$$
$$\mathbf{511,725} \div 5 = 102,345$$
$$\mathbf{9,876,630} \div 5 = 1,975,326$$

Divisible by 10, 100, or 1,000

Every number that ends in 0 is divisible by 10. The following bolded numbers are divisible by 10:

$$\mathbf{20} \div 10 = 2$$
$$\mathbf{170} \div 10 = 17$$
$$\mathbf{56,720} \div 10 = 5,672$$

Every number that ends in 00 is divisible by 100:

$$\mathbf{300} \div 100 = 3$$
$$\mathbf{8,300} \div 100 = 83$$
$$\mathbf{634,900} \div 100 = 6,349$$

And every number that ends in 000 is divisible by 1,000:

$$\mathbf{6,000} \div 1,000 = 6$$
$$\mathbf{99,000} \div 1,000 = 99$$
$$\mathbf{1,234,000} \div 1,000 = 1,234$$

In general, every number that ends with a string of 0s is divisible by the number you get when you write 1 followed by that many 0s. For example,

900,000 is divisible by 100,000.

235,000,000 is divisible by 1,000,000.

820,000,000,000 is divisible by 10,000,000,000.

When numbers start to get this large, mathematicians usually switch over to *scientific notation* to write them more efficiently. In Chapter 14, I show you how to work with scientific notation.

Add it up: Checking divisibility by adding up digits

Sometimes you can check divisibility by adding up all or some of the digits in a number. The sum of a number's digits is called its *digital root.* Finding the digital root of a number is easy, and it's handy to know.

To find the digital root of a number, just add up the digits and repeat this process until you get a one-digit number. Here are some examples:

The digital root of 24 is 6 because $2 + 4 = 6$.

The digital root of 143 is 8 because $1 + 4 + 3 = 8$.

The digital root of 51,111 is 9 because $5 + 1 + 1 + 1 + 1 = 9$.

Sometimes you need to do this process more than once. Here's how to find the digital root of the number 87,482. You have to repeat the process three times, but eventually you find that the digital root of 87,482 is 2:

$8 + 7 + 4 + 8 + 2 = 29$

$2 + 9 = 11$

$1 + 1 = 2$

Read on to find out how sums of digits can help you check for divisibility by 3, 9, or 11.

Divisible by 3

Every number whose digital root is 3, 6, or 9 is divisible by 3.

First, find the digital root of a number by adding its digits until you get a single-digit number. Here are the digital roots of 18, 51, and 975:

18: $1 + 8 = 9$
51: $5 + 1 = 6$
975: $9 + 7 + 5 = 21; 2 + 1 = 3$

With the numbers 18 and 51, adding the digits leads immediately to digital roots 9 and 6, respectively. With 975, when you add up the digits, you first get 21, so you then add up the digits in 21 to get the digital root 3. Thus, these three numbers are all divisible by 3. If you do the actual division, you find that $18 \div 3 = 6$, $51 \div 3 = 17$, and $975 \div 3 = 325$, so the method checks out.

However, when the digital root of a number is anything other than 3, 6, or 9, the number *isn't* divisible by 3:

1,037: $1 + 0 + 3 + 7 = 11; 1 + 1 = 2$

Because the digital root of 1,037 is 2, 1,037 *isn't* divisible by 3. If you try to divide by 3, you end up with 345r2.

Divisible by 9

Every number whose digital root is 9 is divisible by 9.

To test whether a number is divisible by 9, find its digital root by adding up its digits until you get a one-digit number. Here are some examples:

36: $3 + 6 = 9$
243: $2 + 4 + 3 = 9$
7,587: $7 + 5 + 8 + 7 = 27; 2 + 7 = 9$

With the numbers 36 and 243, adding the digits leads immediately to digital roots of 9 in both cases. With 7,587, however, when you add up the digits, you get 27, so you then add up the digits in 27 to get the digital root 9. Thus, all three of these numbers are divisible by 9. You can verify this by doing the division:

$36 \div 9 = 4 \quad 243 \div 9 = 27 \quad 7{,}857 \div 9 = 873$

However, when the digital root of a number is anything other than 9, the number isn't divisible by 9. Here's an example:

706: $7 + 0 + 6 = 13; 1 + 3 = 4$

Because the digital root of 706 is 4, 706 *isn't* divisible by 9. If you try to divide 706 by 9, you get 78r4.

Ups and downs: Divisibility by 11

Two-digit numbers that are divisible by 11 are hard to miss because they simply repeat the same digit twice. Here are all the numbers less than 100 that are divisible by 11:

11 22 33 44 55 66 77 88 99

For numbers between 100 and 200, use this rule: Every three-digit number whose first and third digits add up to its second digit is divisible by 11. For example, suppose you want to decide whether the number 154 is divisible by 11. Just add the first and third digits:

$1 + 4 = 5$

Because these two numbers add up to the second digit, 5, the number 154 is divisible by 11. If you divide, you get $154 \div 11 = 14$, a whole number.

Now suppose you want to figure out whether 136 is divisible by 11. Add the first and third digits:

$1 + 6 = 7$

Because the first and third digits add up to 7 instead of 3, the number 136 isn't divisible by 11. You can find that $136 \div 11 = 12r4$.

For numbers of any length, the rule is slightly more complicated, but it's still often easier than doing long division. To find out when a number is divisible by 11, place plus and minus signs alternatively in front of every digit, then calculate the result. If this result is divisible by 11 (including 0), the number is divisible by 11; otherwise, the number isn't divisible by 11.

For example, suppose you want to discover whether the number 15,983 is divisible by 11. To start out, place plus and minus signs in front of alternate digits (every other digit):

$+1 - 5 + 9 - 8 + 3 = 0$

Because the result is 0, the number 15,983 is divisible by 11. If you check the division, $15{,}983 \div 11 = 1{,}453$.

Now suppose you want to find out whether 9,181,909 is divisible by 11. Again, place plus and minus signs in front of alternate digits and calculate the result:

$+9 - 1 + 8 - 1 + 9 - 0 + 9 = 33$

Because 33 is divisible by 11, the number 9,181,909 is also divisible by 11. The actual answer is

$9{,}181{,}909 \div 11 = 834{,}719$

Identifying Prime and Composite Numbers

In the earlier section titled "Counting everyone in: Numbers you can divide everything by," I show you that every number (except 0 and 1) is divisible by at least two numbers: 1 and itself. In this section, I explore prime numbers and composite numbers (which I introduce you to in Chapter 1).

In Chapter 8, you need to know how to tell prime numbers from composite to break a number down into its prime factors. This tactic is important when you begin working with fractions.

A *prime number* is divisible by exactly two positive whole numbers: 1 and the number itself. A *composite number* is divisible by at least three numbers.

For example, 2 is a prime number because when you divide it by any number but 1 and 2, you get a remainder. So there's only one way to multiply two counting numbers and get 2 as a product:

$$1 \times 2 = 2$$

Similarly, 3 is prime because when you divide by any number but 1 or 3, you get a remainder. So the only way to multiply two numbers together and get 3 as a product is the following:

$$1 \times 3 = 3$$

On the other hand, 4 is a composite number because it's divisible by three numbers: 1, 2, and 4. In this case, you have two ways to multiply two counting numbers and get a product of 4:

$$1 \times 4 = 4$$
$$2 \times 2 = 4$$

But 5 is a prime number because it's divisible only by 1 and 5. Here's the only way to multiply two counting numbers and get 5 as a product:

$$1 \times 5 = 5$$

And 6 is a composite number because it's divisible by 1, 2, 3, and 6. Here are two ways to multiply two counting numbers and get a product of 6:

$1 \times 6 = 6$

$2 \times 3 = 6$

Every counting number except 1 is either prime or composite. The reason 1 is neither is that it's divisible by only *one* number, which is 1.

Here's a list of the prime numbers that are less than 30:

2, 3, 5, 7, 11, 13, 17, 19, 23, 29

Remember the first four prime numbers: 2, 3, 5, and 7. Every composite number less than 100 is divisible by at least one of these numbers. This fact makes it easy to test whether a number under 100 is prime: Simply test it for divisibility by 2, 3, 5, and 7. If it's divisible by any of these numbers, it's composite — if not, it's prime.

For example, suppose you want to find out whether the number 79 is prime or composite without actually doing the division. Here's how you think it out, using the tricks I show you earlier in "Knowing the Divisibility Tricks":

- 79 is an odd number, so it isn't divisible by 2.
- 79 has a digital root of 7 (because $7 + 9 = 16$; $1 + 6 = 7$), so it isn't divisible by 3.
- 79 doesn't end in 5 or 0, so it isn't divisible by 5.
- Even though there's no trick for divisibility by 7, you know that 77 is divisible by 7. So $79 \div 7$ leaves a remainder of 2, which tells you that 79 isn't divisible by 7.

Because 79 is less than 100 and isn't divisible by 2, 3, 5, or 7, you know that 79 is a prime number.

Now test whether 93 is prime or composite:

- 93 is an odd number, so it isn't divisible by 2.
- 93 has a digital root of 3 (because $9 + 3 = 12$ and $1 + 2 = 3$), so 93 is divisible by 3.

You don't need to look further. Because 93 is divisible by 3, you know it's composite.

Chapter 8

Fabulous Factors and Marvelous Multiples

In This Chapter

- Understanding how factors and multiples are related
- Listing all the factors of a number
- Breaking down a number into its prime factors
- Generating multiples of a number
- Finding the greatest common factor (GCF) and least common multiple (LCM)

In Chapter 2, I introduce you to sequences of numbers based on the multiplication table. In this chapter, I tell you about two important ways to think about these sequences: as *factors* and as *multiples.* Factors and multiples are really two sides of the same coin. Here I show you what you need to know about these two important concepts.

For starters, I discuss how factors and multiples are connected to multiplication and division. Then I show you how to find all the factor pairs of a number and how to decompose (split up) any number into its prime factors. To finish up on factors, I show you how to find the greatest common factor (GCF) of any set of numbers. After that, I tackle multiples, showing you how to generate the multiples of a number and then use this skill to find the least common multiple (LCM) of a set of numbers.

Knowing Six Ways to Say the Same Thing

In this section, I introduce you to factors and multiples, and I show you how these two important concepts are connected. As I discuss in Chapter 4, multiplication and division are inverse operations. For example, the following equation is true:

$$5 \times 4 - 20$$

So this equation using the inverse operation is also true:

$$20 \div 4 = 5$$

You may have noticed that, in math, you tend to run into the same ideas over and over again. For example, mathematicians have six different ways to talk about this relationship.

The following three statements all focus on the relationship between 5 and 20 from the perspective of multiplication:

- 5 *multiplied* by some number is 20.
- 5 is a *factor* of 20.
- 20 is a *multiple* of 5.

In two of the examples, you can see this relationship reflected in the words *multiplied* and *multiple.* For the remaining example, keep in mind that two factors are multiplied to equal a product.

Similarly, the following three statements all focus on the relationship between 5 and 20 from the perspective of division:

- 20 *divided* by some number is 5.
- 20 is *divisible by* 5.
- 5 is a *divisor* of 20.

Why do mathematicians need all these words for the same thing? Maybe for the same reason that Eskimos need a bunch of words for snow. In any case, in this chapter, I focus on the words *factor* and *multiple.* When you understand the concepts, which word you choose doesn't matter a whole lot.

Connecting Factors and Multiples

When one number is a factor of a second number, the second number is a multiple of the first number. For example, 20 is divisible by 5, so

- 5 is a factor of 20.
- 20 is a multiple of 5.

Don't mix which number is the factor and which is the multiple. The factor is always the smaller number, and the multiple is always the larger number for positive numbers.

If you have trouble remembering which number is the factor and which is the multiple, jot them down in order from lowest to highest, and write the letters F and M in alphabetical order under them.

For example, 10 divides 40 evenly, so jot down:

10	***40***
F	M

This setup should remind you that 10 is a factor of 40 and that 40 is a multiple of 10.

Finding Fabulous Factors

In this section, I introduce you to factors. First, I show you how to find out whether one number is a factor of another. Then I show you how to list all the factor pairs of a number. After that, I introduce the key idea of a number's prime factors. This information all leads up to an essential skill: finding the greatest common factor (GCF) of a set of numbers.

Deciding when one number is a factor of another

You can easily tell whether a number is a factor of a second number: Just divide the second number by the first. If it divides evenly (with no remainder), the number is a factor; otherwise, it's not a factor.

For example, suppose you want to know whether 7 is a factor of 56. Here's how you find out:

$$56 \div 7 = 8$$

Because 7 divides 56 without leaving a remainder, 7 is a factor of 56.

And here's how you find out whether 4 is a factor of 34:

$$34 \div 4 = 8r2$$

Because 4 divides 34 with a remainder of 2, 4 isn't a factor of 34.

This method works no matter how large the numbers are.

Some teachers use factoring problems to test you on long division. For a refresher on how to do long division, see Chapter 3.

Understanding factor pairs

A factor pair of a number is any pair of two numbers that, when multiplied together, equal that number. For example, 35 has two factor pairs — 1×35 and 5×7 because

$$1 \times 35 = 35$$
$$5 \times 7 = 35$$

Similarly, 24 has four factor pairs — 1×24, 2×12, 3×8, and 4×6 — because

$$1 = 24$$
$$2 \times 12 = 24$$
$$3 \times 8 = 24$$
$$4 \times 6 = 24$$

Every positive integer has at least one factor pair: 1 times the number itself. For example:

$$1 \times 2 = 2 \quad 1 \times 11 = 11 \quad 1 \times 43 = 43$$

When a number greater than 1 has only one factor pair, it's a prime number (see Chapter 7 for more on prime numbers).

Generating a number's factors

The greatest factor of any number is the number itself, so you can always list all the factors of any number because you have a stopping point. A good way to list all the factors of a number is to list all its factor pairs:

1. **Begin the list with 1 times the number itself.**
2. **Try to find a factor pair that includes 2.**

   That is, see whether the number is divisible by 2 (for tricks on testing for divisibility, see Chapter 7). If it is, list the factor pair that includes 2.
3. **Test the number 3 in the same way.**
4. **Continue testing numbers until you find no more factor pairs.**

An example can help make this clear. Suppose you want to list all the factors of the number 18. According to Step 1, begin with 1×18:

1×18

Remember from Chapter 7 that every number — whether prime or composite — is divisible by itself and 1. So automatically, 1 and 18 are both factors of 18.

Next, see if you can find a factor pair of 18 that includes 2. Of course, 18 is an even number, so you know that such a factor pair exists. (For a bunch of easy divisibility tricks, check out Chapter 3.) Here it is:

2×9

Because 2 divides 18 without a remainder, 2 is a factor of 18. (For a bunch of easy divisibility tricks, check out Chapter 3.) So both 2 *and* 9 are factors of 18, and you can add them both to the list:

Now test 3 in the same way:

3×6

At this point, you're almost done. You have to check only the numbers between 3 and 6 — that is, the numbers 4 and 5. Neither of these numbers is included in a factor pair of 18 because 18 isn't divisible by 4 or 5:

$18 \div 4 = 4r2$

$18 \div 5 = 3r3$

So 18 has three factor pairs — 1×18, 2×9, and 3×6 — and thus has six factors. If you like (or if your teacher prefers!), you can list these factors in order, as follows:

1 2 3 6 9 18

Identifying prime factors

In Chapter 7, I discuss prime numbers and composite numbers. A *prime number* is divisible only by 1 and itself — for example, the number 7 is divisible only by 1 and 7. On the other hand, a *composite number* is divisible by at least one number other than 1 and itself — for example, the number 9 is divisible not only by 1 and 9, but also by 3.

A number's *prime factors* are the set of prime numbers (including repeats) that equal that number when multiplied together. For example, here are the prime factors of the numbers 10, 30, and 72:

$$10 = 2 \times 5$$
$$30 = 2 \times 3 \times 5$$
$$72 = 2 \times 2 \times 2 \times 3 \times 3$$

In the last example, the prime factors of 72 include the number 2 repeated three times and the number 3 repeated twice.

The best way to break down a composite number into its prime factors is to use a factorization tree. Here's how it works:

1. **Split the number into any two factors and check off the original number.**
2. **If either of these factors is prime, circle it.**
3. **Repeat Steps 1 and 2 for any number that is neither circled nor checked.**
4. **When every number in the tree is either checked or circled, the tree is finished, and the circled numbers are the prime factors of the original number.**

For example, to break down the number 56 into its prime factors, start by finding two numbers (other than 1 or 56) that, when multiplied, give you a product of 56. In this case, remember that $7 \times 8 = 56$. See Figure 8-1.

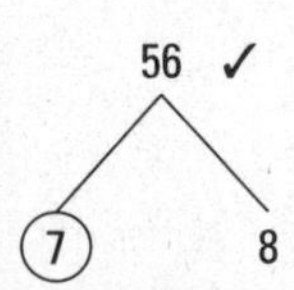

Figure 8-1: Finding two factors of 56; 7 is prime.

Illustration by Wiley, Composition Services Graphics

As you can see, I break down 56 into two factors and check it off. I also circle 7 because it's a prime number. Now, 8 is a neither checked nor circled, so I repeat the process, as shown in Figure 8-2.

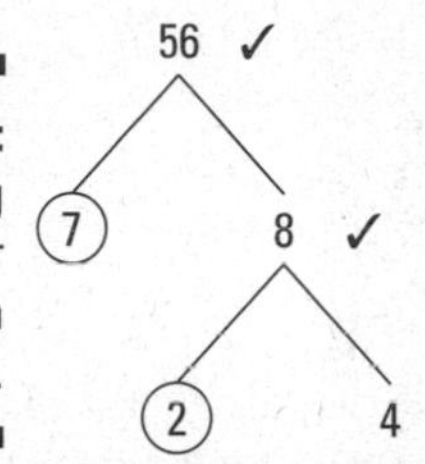

Figure 8-2: Continuing the number breakdown with 8.

Illustration by Wiley, Composition Services Graphics

This time, I break 8 into two factors ($2 \times 4 = 8$) and check it off. This time, 2 is prime, so I circle it. But 4 remains unchecked and uncircled, so I continue with Figure 8-3.

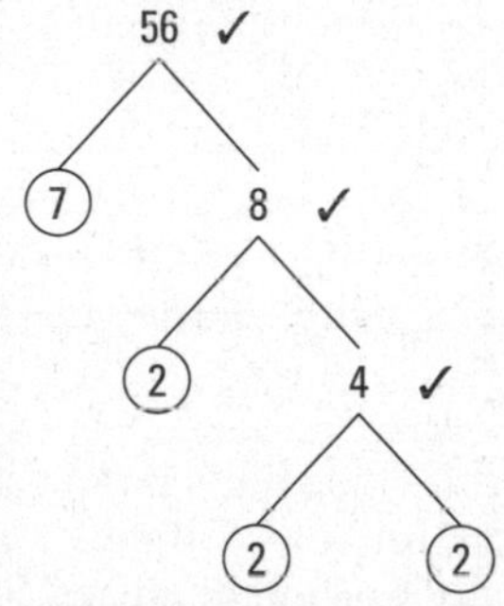

Figure 8-3: The finished tree, completed from Figure 8-1.

Illustration by Wiley, Composition Services Graphics

At this point, every number in the tree is either circled or checked, so the tree is finished. The four circled numbers — 2, 2, 2, and 7 — are the prime factors of 56. To check this result, just multiply the prime factors:

$$2 \times 2 \times 2 \times 7 = 56$$

You can see why this is called a tree: Starting at the top, the numbers tend to branch off like an upside-down tree.

What happens when you try to build a tree starting with a prime number — for example, 7? Well, you don't have to go very far (see Figure 8-4).

Figure 8-4: Starting with a prime number.

Illustration by Wiley, Composition Services Graphics

That's it — you're done! This example shows you that every prime number is its own prime factor.

Here's a list of numbers less than 20 with their prime factorizations. (As you find out in Chapter 2, 1 is neither prime nor composite, so it doesn't have a prime factorization.)

$2 = 2$	$8 = 2 \times 2 \times 2$	$14 = 2 \times 7$
$3 = 3$	$9 = 3 \times 3$	$15 = 3 \times 5$
$4 = 2 \times 2$	$10 = 2 \times 5$	$16 = 2 \times 2 \times 2 \times 2$
$5 = 5$	$11 = 11$	$17 = 17$
$6 = 2 \times 3$	$12 = 2 \times 2 \times 3$	$18 = 2 \times 3 \times 3$
$7 = 7$	$13 = 13$	$19 = 19$

As you can see, the eight prime numbers that I list here are their own prime factorizations. The remaining numbers are composite, so they can all be broken down into smaller prime factors.

Every number has a unique prime factorization. This fact is important — so important that it's called the Fundamental Theorem of Arithmetic. In a way, a number's prime factorization is like its fingerprint — a unique and foolproof way to identify a number.

Knowing how to break down a number to its prime factorization is a handy skill to have. Using the factorization tree allows you to factor out one number after another until all you're left with are primes.

Finding prime factorizations for numbers 100 or less

When you build a factorization tree, the first step is usually the hardest. That's because, as you proceed, the numbers get smaller and easier to work with. With fairly small numbers, the factorization tree is usually easy to use.

As the number you're trying to factor increases, you may find the first step to be a little more difficult. It's especially hard when you don't recognize the number from the multiplication table. The trick is to find someplace to start.

Whenever possible, factor out 5s and 2s first. As I discuss in Chapter 7, you can easily tell when a number is divisible by 2 or by 5.

For example, suppose you want the prime factorization of the number 84. Because you know that 84 is divisible by 2, you can factor out a 2, as shown in Figure 8-5.

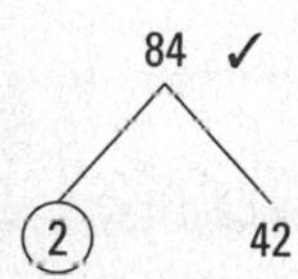

Figure 8-5: Factoring out 2 from 84.

Illustration by Wiley, Composition Services Graphics

At this point, you should recognize 42 from the multiplication table ($6 \times 7 - 42$).

This tree is now easy to complete (see Figure 8-6).

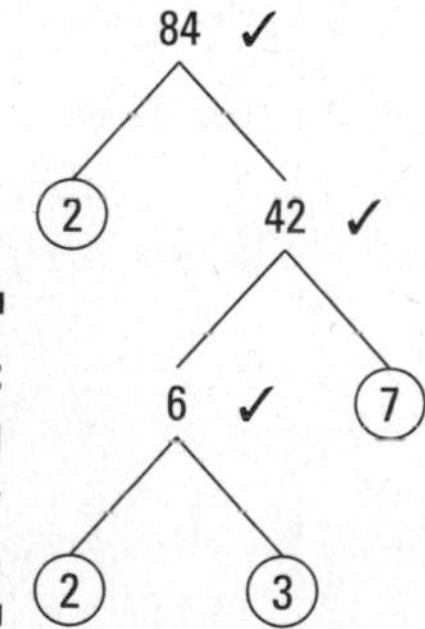

Figure 8-6: Completing the factoring of 84.

Illustration by Wiley, Composition Services Graphics

The resulting prime factorization for 84 is as follows:

$$84 = 2 \times 7 \times 2 \times 3$$

If you like, though, you can rearrange the factors from lowest to highest:

$$84 = 2 \times 2 \times 3 \times 7$$

By far, the most difficult situation occurs when you're trying to find the prime factors of a prime number but don't know it. For example, suppose you want to find the prime factorization for the number 71. This time, you don't recognize the number from the multiplication tables, and it isn't divisible by 2 or 5. What next?

If a number that's less than 100 (actually, less than 121) isn't divisible by 2, 3, 5, or 7, it's a prime number.

Testing for divisibility by 3 by finding the digital root of 71 (that is, by adding the digits) is easy. As I explain in Chapter 7, numbers divisible by 3 have digital roots of 3, 6, or 9.

$$7 + 1 = 8$$

Because the digital root of 71 is 8, 71 isn't divisible by 3. Divide to test whether 71 is divisible by 7:

$$71 \div 7 = 10r1$$

So now you know that 71 isn't divisible by 2, 3, 5, or 7. Therefore, 71 is a prime number, so you're done.

Finding prime factorizations for numbers greater than 100

Most of the time, you don't have to worry about finding the prime factorizations of numbers greater than 100. Just in case, though, here's what you need to know.

As I mention in the preceding section, factor out the 5s and 2s first. A special case is when the number you're factoring ends in one or more 0s. In this case, you can factor out a 10 for every 0. For example, Figure 8-7 shows the first step.

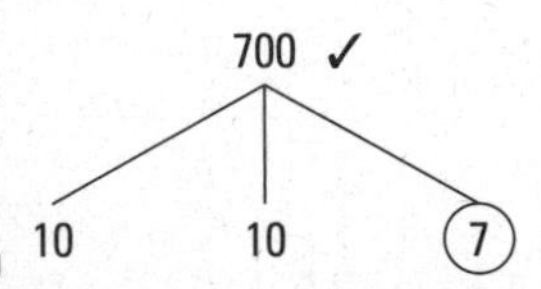

Figure 8-7: The first step in factoring 700.

Illustration by Wiley, Composition Services Graphics

After you do the first step, the rest of the tree becomes easy (see Figure 8-8):

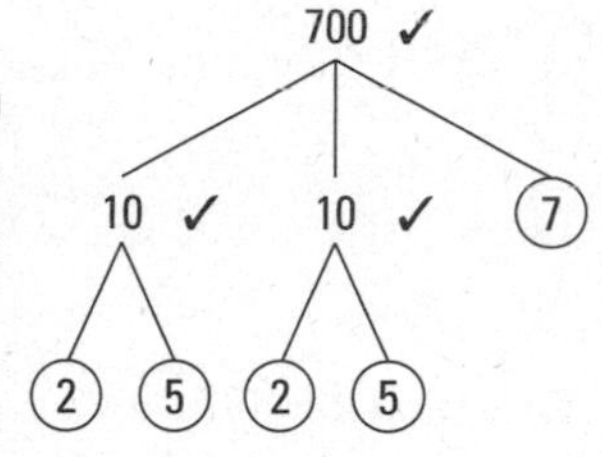

Figure 8-8: Completing the factoring of 700.

Illustration by Wiley, Composition Services Graphics

You can see that the prime factorization of 700 is

$$700 = 2 \times 2 \times 5 \times 5 \times 7$$

If the number isn't divisible by either 2 or 5, use your divisibility trick for 3 (see Chapter 7) and factor out as many 3s as you can. Then factor out 7s, if possible (sorry, I don't have a trick for 7s), and, finally, 11s.

If a number that's less than 289 isn't divisible by 2, 3, 5, 7, 11, or 13, it's prime. As always, every prime number is its own prime factorization, so when you know that a number is prime, you're done. Most of the time, with larger numbers, a combination of tricks can handle the job.

Finding the greatest common factor (GCF)

When you understand how to find the factors of a number (see "Generating a number's factors"), you're ready to move on to the main event: finding the greatest common factor of several numbers.

The greatest common factor (GCF) of a set of numbers is the largest number that's a factor of all those numbers. For example, the GCF of the numbers 4 and 6 is 2 because 2 is the greatest number that's a factor of both 4 and 6.

To find the GCF of a set of numbers, list all the factors of each number, as I show you in "Generating a number's factors." The greatest factor appearing on every list is the GCF. For example, to find the GCF of 6 and 15, first list all the factors of each number.

Factors of 6: 1, 2, 3, 6

Factors of 15: 1, 3, 5, 15

Because 3 is the greatest factor that appears on both lists, 3 is the GCF of 6 and 15.

As another example, suppose you want to find the GCF of 9, 20, and 25. Start by listing the factors of each:

Factors of 9: 1, 3, 9

Factors of 20: 1, 2, 4, 5, 10, 20

Factors of 25: 1, 5, 25

In this case, the only factor that appears on all three lists is 1, so 1 is the GCF of 9, 20, and 25.

Making Marvelous Multiples

Even though multiples tend to be larger numbers than factors, most students find them easier to work with. Read on.

Generating multiples

The preceding section, "Finding Fabulous Factors," tells you how to find all the factors of a number. Finding all the factors is possible because factors of a number are always less than or equal to the number itself. So no matter how large a number is, it always has a finite (limited) number of factors.

Unlike factors, multiples of a number are greater than or equal to the number itself. (The only exception to this is 0, which is a multiple of every number.) Because of this, the multiples of a number go on forever — that is, they're infinite. Nevertheless, generating a partial list of multiples for any number is simple.

To list multiples of any number, write down that number and then multiply it by 2, 3, 4, and so forth.

For example, here are the first few positive multiples of 7:

7 14 21 28 35 42

As you can see, this list of multiples is simply part of the multiplication table for the number 7. (For the multiplication table up to 9 × 9, see Chapter 3.)

Finding the least common multiple (LCM)

The least common multiple (LCM) of a set of numbers is the lowest positive number that's a multiple of every number in that set.

For example, the LCM of the numbers 2, 3, and 5 is 30 because

- 30 is a multiple of 2 (2 × 15 = 30).
- 30 is a multiple of 3 (3 × 10 = 30).
- 30 is a multiple of 5 (5 × 6 = 30).
- No number lower than 30 is a multiple of all three numbers.

To find the LCM of a set of numbers, take each number in the set and jot down a list of the first several multiples in order. The LCM is the first number that appears on every list.

When looking for the LCM of two numbers, start by listing multiples of the larger number, but stop this list when the number of multiples you've written equals the smaller number. Then start listing multiples of the lower number until one of them matches the first list.

For example, suppose you want to find the LCM of 4 and 6. Begin by listing multiples of the higher number, which is 6. In this case, list only four of these multiples because the lower number is 4.

Multiples of 6: 6, 12, 18, 24, …

Now start listing multiples of 4:

Multiples of 4: 4, 8, 12, …

Because 12 is the first number to appear on both lists of multiples, 12 is the LCM of 4 and 6.

This method works especially well when you want to find the LCM of two numbers, but it may take longer if you have more numbers.

Suppose, for instance, that you want to find the LCM of 2, 3, and 5. Start with the two largest numbers — in this case, 5 and 3 — and keep listing numbers until you have one or more matching numbers:

Multiples of 5: 5, 10, 15, 20, 25, 30, ...

Multiples of 3: 3, 6, 9, 12, 15, 18, 21, 24, 27, 30, ...

The only numbers repeated on both lists are 15 and 30. In this case, you can save yourself the trouble of making the last list because 30 is obviously a multiple of 2, and 15 isn't. So 30 is the LCM of 2, 3, and 5.

Part III

Parts of the Whole: Fractions, Decimals, and Percents

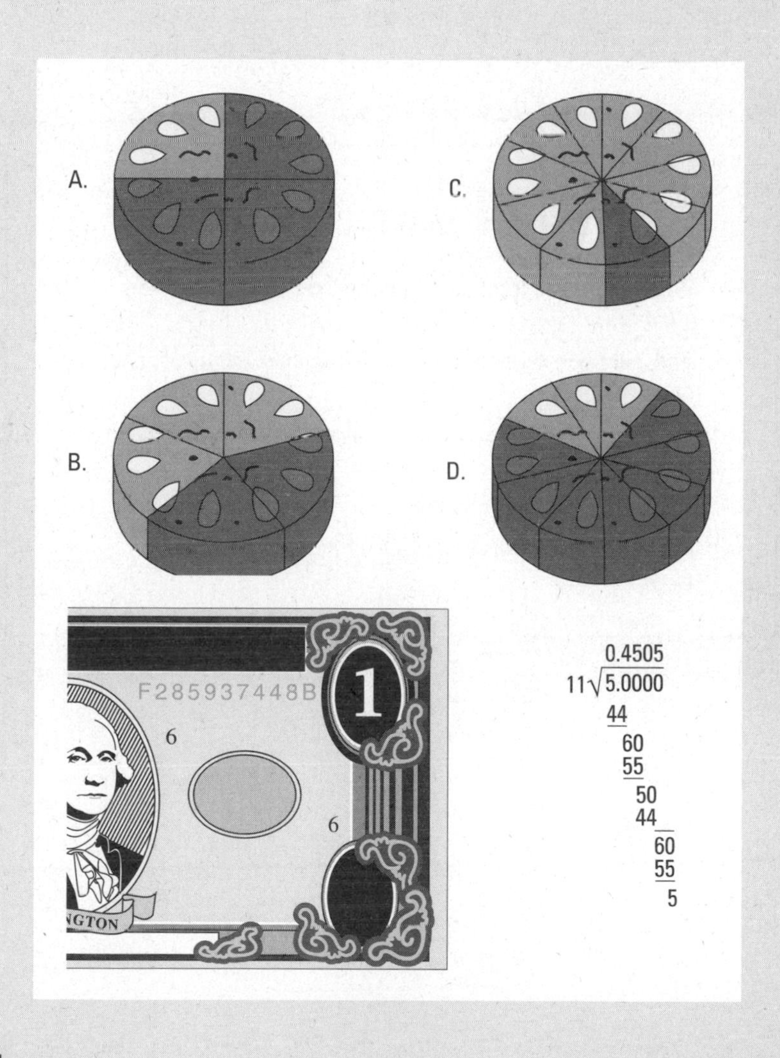

To discover a great way to solve percent problems, go to www.dummies.com/extras/basicmathandprealgebra.

In this part...

- Work with basic fractions, improper fractions, and mixed numbers
- Add, subtract, multiply, and divide fractions, decimals, and percents
- Convert the form of a rational number to a fraction, a decimal, or a percent
- Use ratios and proportions
- Solve word problems that involve fractions, decimals, percentages

Chapter 9

Fooling with Fractions

In This Chapter

- Looking at basic fractions
- Knowing the numerator from the denominator
- Understanding proper fractions, improper fractions, and mixed numbers
- Increasing and reducing the terms of fractions
- Converting between improper fractions and mixed numbers
- Using cross-multiplication to compare fractions

Suppose that today is your birthday and your friends are throwing you a surprise party. After opening all your presents, you finish blowing out the candles on your cake, but now you have a problem: Eight of you want some cake, but you have only *one cake*. Several solutions are proposed:

- You can all go into the kitchen and bake seven more cakes.
- Instead of eating cake, everyone can eat celery sticks.
- Because it's your birthday, you can eat the *whole* cake and everyone else can eat celery sticks. (That idea was yours.)
- You can cut the cake into eight equal slices so that everyone can enjoy it.

After careful consideration, you choose the last option. With that decision, you've opened the door to the exciting world of fractions. Fractions represent parts of a thing that can be cut into pieces. In this chapter, I give you some basic information about fractions that you need to know, including the three basic types of fractions: proper fractions, improper fractions, and mixed numbers.

I move on to increasing and reducing the terms of fractions, which you need when you begin applying the Big Four operations to fractions in Chapter 10. I also show you how to convert between improper fractions and mixed numbers. Finally, I show you how to compare fractions using cross-multiplication. By the time you're done with this chapter, you'll see how fractions really can be a piece of cake!

Slicing a Cake into Fractions

Here's a simple fact: When you cut a cake into two equal pieces, each piece is half of the cake. As a fraction, you write that as $\frac{1}{2}$. In Figure 9-1, the shaded piece is half of the cake.

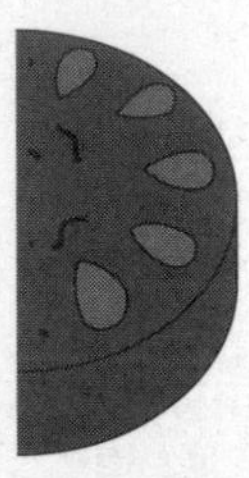

Figure 9-1: Two halves of a cake.

Illustration by Wiley, Composition Services Graphics

Every fraction is made up of two numbers separated by a line, or a fraction bar. The line can be either diagonal or horizontal — so you can write this fraction in either of the following two ways:

$\frac{1}{2}$ 1/2

The number above the line is called the numerator. The numerator tells you how many pieces you have. In this case, you have one dark-shaded piece of cake, so the numerator is 1.

The number below the line is called the denominator. The denominator tells you how many equal pieces the whole cake has been cut into. In this case, the denominator is 2.

Similarly, when you cut a cake into three equal slices, each piece is a third of the cake (see Figure 9-2).

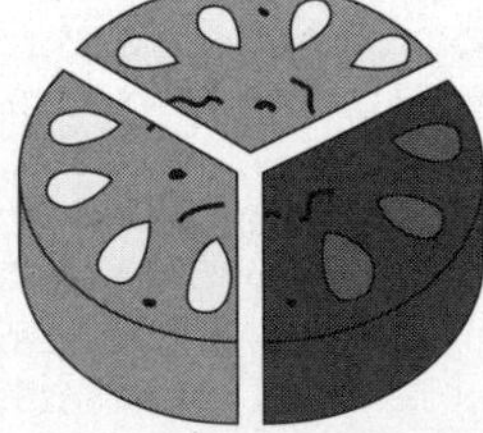

Figure 9-2: Cake cut into thirds.

Illustration by Wiley, Composition Services Graphics

This time, the shaded piece is one-third — $\frac{1}{3}$ — of the cake. Again, the numerator tells you how many pieces you have, and the denominator tells you how many equal pieces the whole cake has been cut up into.

Figure 9-3 shows a few more examples of ways to represent parts of the whole with fractions.

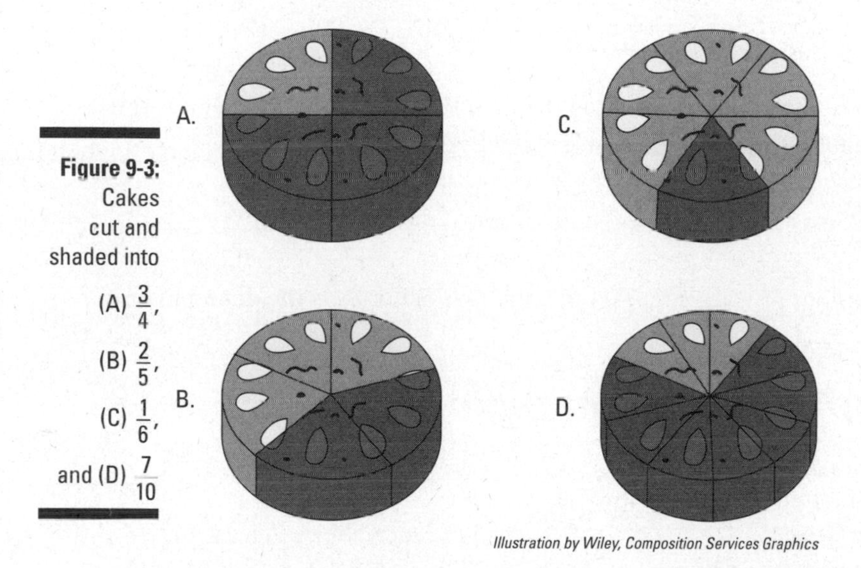

Figure 9-3: Cakes cut and shaded into (A) $\frac{3}{4}$, (B) $\frac{2}{5}$, (C) $\frac{1}{6}$, and (D) $\frac{7}{10}$

Illustration by Wiley, Composition Services Graphics

In each case, the numerator tells you how many pieces are shaded, and the denominator tells how many pieces there are altogether.

The fraction bar can also mean a division sign. In other words, $\frac{3}{4}$ signifies $3 \div 4$. If you take three cakes and divide them among four people, each person gets $\frac{3}{4}$ of a cake.

Knowing the Fraction Facts of Life

Fractions have their own special vocabulary and a few important properties that are worth knowing right from the start. When you know them, you find working with fractions a lot easier.

Telling the numerator from the denominator

The top number in a fraction is called the *numerator,* and the bottom number is called the *denominator.* For example, look at the following fraction:

$$\frac{3}{4}$$

In this example, the number 3 is the numerator, and the number 4 is the denominator. Similarly, look at this fraction:

$$\frac{55}{89}$$

The number 55 is the numerator, and the number 89 is the denominator.

Flipping for reciprocals

When you flip over a fraction, you get its reciprocal. For example, the following numbers are reciprocals:

$$\frac{2}{3} \text{ and } \frac{3}{2}$$

$$\frac{11}{14} \text{ and } \frac{14}{11}$$

$$\frac{19}{19} \text{ is its own reciprocal}$$

Using ones and zeros

When the denominator (bottom number) of a fraction is 1, the fraction is equal to the numerator by itself. Conversely, you can turn any whole number into a fraction by drawing a line and placing the number 1 under it. For example,

$$\frac{2}{1}=2 \quad \frac{9}{1}=9 \quad \frac{157}{1}=157$$

When the numerator and denominator match, the fraction equals 1. After all, if you cut a cake into eight pieces and you keep all eight of them, you have the entire cake. Here are some fractions that equal 1:

$$\frac{8}{8}=1 \quad \frac{11}{11}=1 \quad \frac{365}{365}=1$$

When the numerator of a fraction is 0, the fraction is equal to 0. For example,

$$\frac{0}{1}=0 \quad \frac{0}{12}=0 \quad \frac{0}{113}=0$$

The denominator of a fraction can never be 0. Fractions with 0 in the denominator are *undefined* — that is, they have no mathematical meaning.

Remember from earlier in this chapter that placing a number in the denominator is similar to cutting a cake into that number of pieces. You can cut a cake into two, or ten, or even a million pieces. You can even cut it into one piece (that is, don't cut it at all). But you can't cut a cake into zero pieces. For this reason, putting 0 in the denominator — much like lighting an entire book of matches on fire — is something you should never, never do.

Mixing things up

A mixed number is a combination of a whole number and a proper fraction added together. Here are some examples:

$$1\frac{1}{2} \quad 5\frac{3}{4} \quad 99\frac{44}{100}$$

A mixed number is always equal to the whole number plus the fraction attached to it. So $1\frac{1}{2}$ means $1+\frac{1}{2}$, $5\frac{3}{4}$ means $5+\frac{3}{4}$, and so on.

Knowing proper from improper

When the numerator (top number) is less than the denominator (bottom number), the fraction is less than 1:

$$\frac{1}{2}<1 \quad \frac{3}{5}<1 \quad \frac{63}{73}<1$$

Fractions like these are called are called *proper fractions.* Positive proper fractions are always between 0 and 1. However, when the numerator is greater than the denominator, the fraction is greater than 1. Take a look:

$$\frac{3}{2}>1 \quad \frac{7}{4}>1 \quad \frac{98}{97}>1$$

Any fraction that's greater than 1 is called an *improper fraction.* Converting an improper fraction to a mixed number is customary, especially when it's the final answer to a problem.

An improper fraction is always top heavy, as if it's unstable and wants to fall over. To stabilize it, convert it to a mixed number. Proper fractions are always stable.

Later in this chapter, I discuss improper fractions in more detail when I show you how to convert between improper fractions and mixed numbers.

Increasing and Reducing Terms of Fractions

Take a look at these three fractions:

$$\frac{1}{2} \quad \frac{2}{4} \quad \frac{3}{6}$$

If you cut three cakes (as I do earlier in this chapter) into these three fractions (see Figure 9-4), exactly half of the cake will be shaded, just like in Figure 9-1, no matter how you slice it. (Get it? No matter how you slice it? You may as well laugh at the bad jokes, too — they're free.) The important point here isn't the humor, or the lack of it, but the idea about fractions.

The fractions $\frac{1}{2}$, $\frac{2}{4}$, and $\frac{3}{6}$ are all equal in value. In fact, you can write a lot of fractions that are also equal to these. As long as the numerator is exactly half the denominator, the fractions are all equal to $\frac{1}{2}$ — for example,

$$\frac{11}{22} \quad \frac{100}{200} \quad \frac{1,000,000}{2,000,000}$$

These fractions are equal to $\frac{1}{2}$, but their terms (the numerator and denominator) are different. In this section, I show you how to both increase and reduce the terms of a fraction without changing its value.

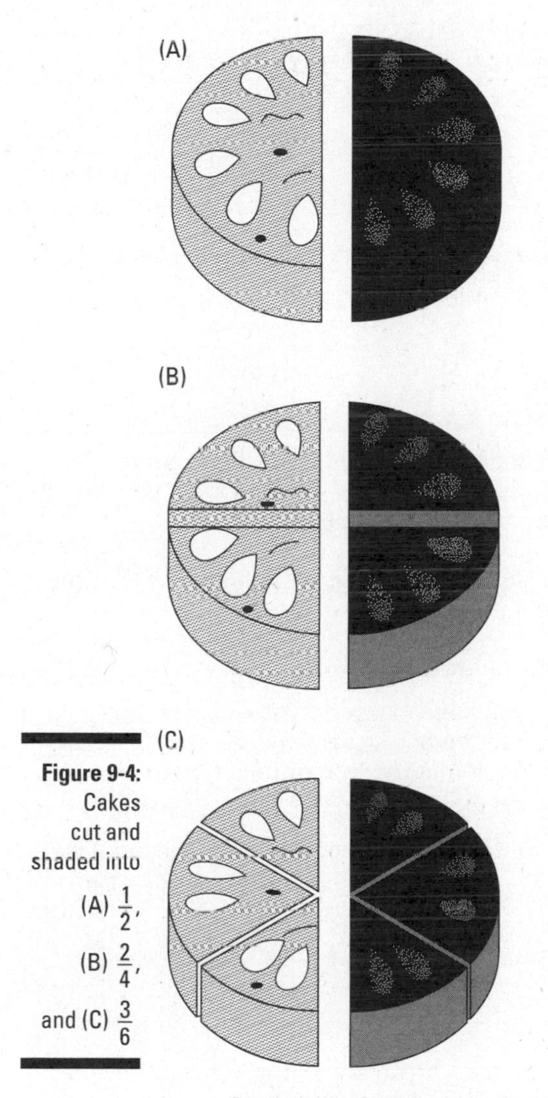

Figure 9-4: Cakes cut and shaded into (A) $\frac{1}{2}$, (B) $\frac{2}{4}$, and (C) $\frac{3}{6}$

Illustration by Wiley, Composition Services Graphics

Increasing the terms of fractions

To increase the terms of a fraction by a certain number, multiply both the numerator and the denominator by that number.

For example, to increase the terms of the fraction $\frac{3}{4}$ by 2, multiply both the numerator and the denominator by 2:

$$\frac{3}{4}=\frac{3\times 2}{4\times 2}=\frac{6}{8}$$

Similarly, to increase the terms of the fraction $\frac{5}{11}$ by 7, multiply both the numerator and the denominator by 7:

$$\frac{5}{11}=\frac{5\times 7}{11\times 7}=\frac{35}{77}$$

Increasing the terms of a fraction doesn't change its value. Because you're multiplying the numerator and denominator by the same number, you're essentially multiplying the fraction by a fraction that equals 1.

One key point to know is how to increase the terms of a fraction so that the denominator becomes a preset number. Here's how you do it:

1. **Divide the new denominator by the old denominator.**

   To keep the fractions equal, you have to multiply the numerator and denominator of the old fraction by the same number. This first step tells you what the old denominator was multiplied by to get the new one.

   For example, suppose you want to raise the terms of the fraction $\frac{4}{7}$ so that the denominator is 35. You're trying to fill in the question mark here:

   $$\frac{4}{7}=\frac{?}{35}$$

   Divide 35 by 7, which tells you that the denominator was multiplied by 5.

2. **Multiply this result by the old numerator to get the new numerator.**

   You now know how the two denominators are related. The numerators need to have the same relationship, so multiply the old numerator by the number you found in Step 1.

   Multiply 5 by 4, which gives you 20. So here's the answer:

   $$\frac{4}{7}=\frac{4\times 5}{7\times 5}=\frac{20}{35}$$

Reducing fractions to lowest terms

Reducing fractions is similar to increasing fractions, except that it involves division rather than multiplication. But because you can't always divide, reducing takes a bit more finesse.

In practice, reducing fractions is similar to factoring numbers. For this reason, if you're not up on factoring, you may want to review this topic in Chapter 8.

In this section, I show you the formal way to reduce fractions, which works in all cases. Then I show you a more informal way you can use when you're more comfortable.

Reducing fractions the formal way

Reducing fractions the formal way relies on understanding how to break down a number into its prime factors. I discuss this in detail in Chapter 8, so if you're shaky on this concept, you may want to review it first.

Here's how to reduce a fraction:

1. **Break down both the numerator (top number) and the denominator (bottom number) into their prime factors.**

   For example, suppose you want to reduce the fraction $\frac{12}{30}$. Break down both 12 and 30 into their prime factors:

   $$\frac{12}{30} = \frac{2 \times 2 \times 3}{2 \times 3 \times 5}$$

2. **Cross out any common factors.**

   As you can see, I cross out a 2 and a 3 because they're common factors — that is, they appear in both the numerator and the denominator:

   $$\frac{12}{30} = \frac{\cancel{2} \times 2 \times \cancel{3}}{\cancel{2} \times \cancel{3} \times 5}$$

3. **Multiply the remaining numbers to get the reduced numerator and denominator.**

   You can see now that the fraction $\frac{12}{30}$ reduces to $\frac{2}{5}$:

   $$\frac{12}{30} = \frac{\cancel{2} \times 2 \times \cancel{3}}{\cancel{2} \times \cancel{3} \times 5} = \frac{2}{5}$$

As another example, here's how you reduce the fraction $\frac{32}{100}$:

$$\frac{32}{100}=\frac{\cancel{2}\times\cancel{2}\times2\times2\times2}{\cancel{2}\times\cancel{2}\times5\times5}=\frac{8}{25}$$

This time, cross out two 2s from both the top and the bottom as common factors. The remaining 2s on top and the 5s on the bottom aren't common factors. So the fraction $\frac{32}{100}$ reduces to $\frac{8}{25}$.

Reducing fractions the informal way

Here's an easier way to reduce fractions when you get comfortable with the concept:

1. **If the numerator (top number) and denominator (bottom number) are both divisible by 2 — that is, if they're both even — divide both by 2.**

   For example, suppose you want to reduce the fraction $\frac{36}{60}$. The numerator and the denominator are both even, so divide them both by 2:

   $$\frac{36}{60}=\frac{18}{30}$$

2. **Repeat Step 1 until the numerator or denominator (or both) is no longer divisible by 2.**

   In the resulting fraction, both numbers are still even, so repeat the first step again:

   $$\frac{18}{30}=\frac{9}{15}$$

3. **Repeat Step 1 using the number 3, and then 5, and then 7, continuing testing prime numbers until you're sure that the numerator and denominator have no common factors.**

   Now, the numerator and the denominator are both divisible by 3 (see Chapter 7 for easy ways to tell if one number is divisible by another), so divide both by 3:

   $$\frac{9}{15}=\frac{3}{5}$$

   Neither the numerator nor the denominator is divisible by 3, so this step is complete. At this point, you can move on to test for divisibility by 5, 7, and so on, but you really don't need to. The numerator is 3, and it obviously isn't divisible by any larger number, so you know that the fraction $\frac{36}{60}$ reduces to $\frac{3}{5}$.

Converting between Improper Fractions and Mixed Numbers

In "Knowing the Fraction Facts of Life," I tell you that any fraction whose numerator is greater than its denominator is an improper fraction. Improper fractions are useful and easy to work with, but for some reason, people just don't like them. (The word *improper* should've tipped you off.) Teachers especially don't like them, and they really don't like an improper fraction to appear as the answer to a problem. However, they love mixed numbers. One reason they love them is that estimating the approximate size of a mixed number is easy.

For example, if I tell you to put $\frac{31}{3}$ of a gallon of gasoline in my car, you probably find it hard to estimate roughly how much that is: 5 gallons, 10 gallons, 20 gallons?

But if I tell you to get $10\frac{1}{3}$ gallons of gasoline, you know immediately that this amount is a little more than 10 but less than 11 gallons. Although $10\frac{1}{3}$ is the same as $\frac{31}{3}$, knowing the mixed number is a lot more helpful in practice. For this reason, you often have to convert improper fractions to mixed numbers.

Knowing the parts of a mixed number

Every mixed number has both a whole number part and a fractional part. So the three numbers in a mixed number are

- The whole number
- The numerator
- The denominator

For example, in the mixed number $3\frac{1}{2}$, the whole number part is 3 and the fractional part is $\frac{1}{2}$. So this mixed number is made up of three numbers: the whole number (3), the numerator (1), and the denominator (2). Knowing these three parts of a mixed number is helpful for converting back and forth between mixed numbers and improper fractions.

Converting a mixed number to an improper fraction

To convert a mixed number to an improper fraction, follow these steps:

1. **Multiply the denominator of the fractional part by the whole number, and add the result to the numerator.**

   For example, suppose you want to convert the mixed number $5\frac{2}{3}$ to an improper fraction. First, multiply 3 by 5 and add 2:

   $$3 \times 5 + 2 = 17$$

2. **Use this result as your numerator, and place it over the denominator you already have.**

   Place this result over the denominator:

   $$\frac{17}{3}$$

   So the mixed number $5\frac{2}{3}$ equals the improper fraction $\frac{17}{3}$. This method works for all mixed numbers. Furthermore, if you start with the fractional part reduced, the answer is also reduced (see the earlier "Increasing and Reducing Terms of Fractions" section).

Converting an improper fraction to a mixed number

To convert an improper fraction to a mixed number, divide the numerator by the denominator (see Chapter 3). Then write the mixed number in this way:

- The quotient (answer) is the whole-number part.
- The remainder is the numerator.
- The denominator of the improper fraction is the denominator.

For example, suppose you want to write the improper fraction $\frac{19}{5}$ as a mixed number. First, divide 19 by 5:

$$19 \div 5 = 3\text{r}4$$

Then write the mixed number as follows:

$$3\frac{4}{5}$$

This method works for all improper fractions. And as is true of conversions in the other direction, if you start with a reduced fraction, you don't have to reduce your answer (see "Increasing and Reducing Terms of Fractions").

Understanding Cross-multiplication

Cross-multiplication is a handy little technique to know. You can use it in a few different ways, so I explain it here and then show you an immediate application.

To cross-multiply two fractions, follow these steps:

1. **Multiply the numerator of the first fraction by the denominator of the second fraction and jot down the answer.**
2. **Multiply the numerator of the second fraction by the denominator of the first fraction and jot down the answer.**

For example, suppose you have these two fractions:

$$\frac{2}{3} \quad \frac{4}{7}$$

When you cross-multiply, you get these two numbers:

$$2 \times 7 = 14 \quad 4 \times 3 = 12$$

You can use cross-multiplication to compare fractions and find out which is greater. When you do so, make sure that you start with the numerator of the first fraction.

To find out which of two fractions is larger, cross-multiply and place the two numbers you get, in order, under the two fractions. The larger number is always under the larger fraction. In this case, 14 goes under $\frac{2}{3}$ and 12 goes under $\frac{4}{7}$. The number 14 is greater than 12, so $\frac{2}{3}$ is greater than $\frac{4}{7}$.

For example, suppose you want to find out which of the following three fractions is the greatest:

$$\frac{3}{5} \quad \frac{5}{9} \quad \frac{6}{11}$$

Cross-multiplication works only with two fractions at a time, so pick the first two — $\frac{3}{5}$ and $\frac{5}{9}$ — and then cross-multiply:

$$3 \times 9 = 27 \quad 5 \times 5 = 25$$

Because 27 is greater than 25, you know now that $\frac{3}{5}$ is greater than $\frac{5}{9}$. So you can throw out $\frac{5}{9}$.

Now do the same thing for $\frac{3}{5}$ and $\frac{6}{11}$:

$$3 \times 11 = 33 \quad 5 \times 6 = 30$$

Because 33 is greater than 30, $\frac{3}{5}$ is greater than $\frac{6}{11}$. Pretty straightforward, right? And that set of steps is all you have to know for now. I show you a bunch of great things you can do with this simple skill in the next chapter.

Making Sense of Ratios and Proportions

A *ratio* is a mathematical comparison of two numbers, based on division. For example, suppose you bring 2 scarves and 3 caps with you on a ski vacation. Here are a few ways to express the ratio of scarves to caps:

$$2:3 \quad 2 \text{ to } 3 \quad \frac{2}{3}$$

The simplest way to work with a ratio is to turn it into a fraction. Be sure to keep the order the same: The first number goes on top of the fraction, and the second number goes on the bottom.

In practice, a ratio is most useful when used to set up a *proportion* — that is, an equation involving two ratios. Typically, a proportion looks like a word equation, as follows:

$$\frac{\text{scarves}}{\text{caps}} = \frac{2}{3}$$

For example, suppose you know that both you and your friend Andrew brought the same proportion of scarves to caps. If you also know that Andrew brought 8 scarves, you can use this proportion to find out how many caps he brought. Just increase the terms of the fraction $\frac{2}{3}$ so that the numerator becomes 8. I do this in two steps:

$$\frac{\text{scarves}}{\text{caps}} = \frac{2 \times 4}{3 \times 4}$$

$$\frac{\text{scarves}}{\text{caps}} = \frac{8}{12}$$

As you can see, the ratio 8:12 is equivalent to the ratio 2:3 because the fractions $\frac{2}{3}$ and $\frac{8}{12}$ are equal. Therefore, Andrew brought 12 caps.

Chapter 10

Parting Ways: Fractions and the Big Four Operations

In This Chapter

- Looking at multiplication and division of fractions
- Adding and subtracting fractions in a bunch of different ways
- Applying the four operations to mixed numbers

In this chapter, the focus is on applying the Big Four operations to fractions. I start by showing you how to multiply and divide fractions, which isn't much more difficult than multiplying whole numbers. Surprisingly, adding and subtracting fractions is a bit trickier. I show you a variety of methods, each with its own strengths and weaknesses, and I recommend how to choose which method will work best, depending on the problem you have to solve.

Later in the chapter, I move on to mixed numbers. Again, multiplication and division won't likely pose too much of a problem because the process in each case is almost the same as multiplying and dividing fractions. I save adding and subtracting mixed numbers for the very end. By then, you'll be much more comfortable with fractions and ready to tackle the challenge.

Multiplying and Dividing Fractions

One of the odd little ironies of life is that multiplying and dividing fractions is easier than adding or subtracting them — just two easy steps and you're done! For this reason, I show you how to multiply and divide fractions before I show you how to add or subtract them. In fact, you may find multiplying fractions easier than multiplying whole numbers because the numbers you're working with are usually small. More good news is that dividing fractions is nearly as easy as multiplying them. So I'm not even wishing you good luck — you don't need it!

Multiplying numerators and denominators straight across

Everything in life should be as simple as multiplying fractions. All you need for multiplying fractions is a pen or pencil, something to write on (preferably not your hand), and a basic knowledge of the multiplication table. (See Chapter 3 for a multiplication refresher.)

Here's how to multiply two fractions:

1. **Multiply the numerators (the numbers on top) to get the numerator of the answer.**
2. **Multiply the denominators (the numbers on the bottom) to get the denominator of the answer.**

For example, here's how to multiply $\frac{2}{5} \times \frac{3}{7}$:

$$\frac{2}{5} \times \frac{3}{7} = \frac{2 \times 3}{5 \times 7} = \frac{6}{35}$$

Sometimes when you multiply fractions, you have an opportunity to reduce to lowest terms. (For more on when and how to reduce a fraction, see Chapter 9.) As a rule, math people are crazy about reduced fractions, and teachers sometimes take points off a right answer if you could've reduced it but didn't. Here's a multiplication problem that ends up with an answer that's not in its lowest terms:

$$\frac{4}{5} \times \frac{7}{8} = \frac{4 \times 7}{5 \times 8} = \frac{28}{40}$$

Because the numerator and the denominator are both even numbers, this fraction can be reduced. Start by dividing both numbers by 2:

$$\frac{28 \div 2}{40 \div 2} = \frac{14}{20}$$

Again, the numerator and the denominator are both even, so do it again:

$$\frac{14 \div 2}{20 \div 2} = \frac{7}{10}$$

This fraction is now fully reduced.

When multiplying fractions, you can often make your job easier by canceling out equal factors in the numerator and denominator. Canceling out equal factors makes the numbers that you're multiplying smaller and easier to work with, and it also saves you the trouble of reducing at the end. Here's how it works:

- When the numerator of one fraction and the denominator of the other are the same, change both of these numbers to 1s. (See the nearby sidebar for why this works.)
- When the numerator of one fraction and the denominator of the other are divisible by the same number, factor this number out of both. In other words, divide the numerator and denominator by that common factor. (For more on how to find factors, see Chapter 8.)

One is the easiest number

With fractions, the relationship between the numbers, not the actual numbers themselves, is most important. Understanding how to multiply and divide fractions can give you a deeper understanding of why you can increase or decrease the numbers within a fraction without changing the value of the whole fraction.

When you multiply or divide any number by 1, the answer is the same number. This rule also goes for fractions, so

$$\frac{3}{8} \times 1 = \frac{3}{8} \text{ and } \frac{3}{8} \div 1 = \frac{3}{8}$$

$$\frac{5}{13} \times 1 = \frac{5}{13} \text{ and } \frac{5}{13} \div 1 = \frac{5}{13}$$

$$\frac{67}{70} \times 1 = \frac{67}{70} \text{ and } \frac{67}{70} \div 1 = \frac{67}{70}$$

And as I discuss in Chapter 9, when a fraction has the same number in both the numerator and the denominator, its value is 1. In other words, the fractions $\frac{2}{2}$, $\frac{3}{3}$, and $\frac{4}{4}$ are all equal to 1. Look what happens when you multiply the fraction $\frac{3}{4}$ by $\frac{2}{2}$:

$$\frac{3}{4} \times \frac{2}{2} = \frac{3 \times 2}{4 \times 2} = \frac{6}{8}$$

The net effect is that you've increased the terms of the original fraction by 2. But all you've done is multiply the fraction by 1, so the value of the fraction hasn't changed. The fraction $\frac{6}{8}$ is equal to $\frac{3}{4}$.

Similarly, reducing the fraction $\frac{6}{9}$ by a factor of 3 is the same as dividing that fraction by $\frac{3}{3}$ (which is equal to 1):

$$\frac{6}{9} \div \frac{3}{3} = \frac{6 \div 3}{9 \div 3} = \frac{2}{3}$$

So $\frac{6}{9}$ is equal to $\frac{2}{3}$.

For example, suppose you want to multiply the following two numbers:

$$\frac{5}{13} \times \frac{13}{20}$$

You can make this problem easier by canceling out the number 13, as follows:

$$\frac{5}{1\,\cancel{13}} \times \frac{\cancel{13}\;1}{20} = \frac{5 \times 1}{1 \times 20} = \frac{5}{20}$$

You can make it even easier by noticing that 20 and 5 are both divisible by 5, so you can also factor out the number 5 before multiplying:

$$\frac{1\,\cancel{5}}{1} \times \frac{1}{\cancel{20}\;4} = \frac{1}{1} \times \frac{1}{4} = \frac{1}{4}$$

Doing a flip to divide fractions

Dividing fractions is just as easy as multiplying them. In fact, when you divide fractions, you really turn the problem into multiplication.

To divide one fraction by another, multiply the first fraction by the reciprocal of the second. (As I discuss in Chapter 9, the *reciprocal* of a fraction is simply that fraction turned upside down.)

For example, here's how you turn fraction division into multiplication:

$$\frac{1}{3} \div \frac{4}{5} = \frac{1}{3} \times \mathbf{\frac{5}{4}}$$

As you can see, I turn $\frac{4}{5}$ into its reciprocal — $\frac{5}{4}$ — and change the division sign to a multiplication sign. After that, just multiply the fractions as I describe in "Multiplying numerators and denominators straight across":

$$\frac{1}{3} \times \frac{5}{4} = \frac{1 \times 5}{3 \times 4} = \frac{5}{12}$$

As with multiplication, in some cases, you may have to reduce your result at the end. But you can also make your job easier by canceling out equal factors. (See the preceding section.)

All Together Now: Adding Fractions

When you add fractions, one important item to notice is whether their denominators (the numbers on the bottom) are the same. If they're the same — woo-hoo! Adding fractions that have the same denominator is a walk in the park. But when fractions have different denominators, adding them becomes a tad more complex.

To make matters worse, many teachers make adding fractions even more difficult by requiring you to use a long and complicated method when, in many cases, a short and easy one will do.

In this section, I first show you how to add fractions with the same denominator. Then I show you a foolproof method for adding fractions when the denominators are different. It always works, and it's usually the simplest way to go. After that, I show you a quick method that you can use only for certain problems. Finally, I show you the longer, more complicated way to add fractions that usually gets taught.

Finding the sum of fractions with the same denominator

To add two fractions that have the same denominator (bottom number), add the numerators (top numbers) and leave the denominator unchanged.

For example, consider the following problem:

$$\frac{1}{5}+\frac{2}{5}=\frac{1+2}{5}=\frac{3}{5}$$

As you can see, to add these two fractions, you add the numerators (1 + 2) and keep the denominator (5).

Why does this work? Chapter 9 tells you that you can think about fractions as pieces of cake. The denominator in this case tells you that the entire cake has been cut into five pieces. So when you add $\frac{1}{5}+\frac{2}{5}$, you're really adding one piece plus two pieces. The answer, of course, is three pieces — that is, $\frac{3}{5}$.

Even if you have to add more than two fractions, as long as the denominators are all the same, you just add the numerators and leave the denominator unchanged:

$$\frac{1}{17}+\frac{3}{17}+\frac{4}{17}+\frac{6}{17}=\frac{1+3+4+6}{17}=\frac{14}{17}$$

Sometimes when you add fractions with the same denominator, you have to reduce it to lowest terms (to find out more about reducing, flip to Chapter 9). Take this problem, for example:

$$\frac{1}{4}+\frac{1}{4}=\frac{1+1}{4}=\frac{2}{4}$$

The numerator and the denominator are both even, so you know they can be reduced:

$$\frac{2}{4}=\frac{1}{2}$$

In other cases, the sum of two proper fractions is an improper fraction. You get a numerator that's larger than the denominator when the two fractions add up to more than 1, as in this case:

$$\frac{3}{7}+\frac{5}{7}=\frac{8}{7}$$

If you have more work to do with this fraction, leave it as an improper fraction so that it's easier to work with. But if this is your final answer, you may need to turn it into a mixed number (I cover mixed numbers in Chapter 9):

$$\frac{8}{7}=8\div 7=1\text{r}1=1\frac{1}{7}$$

When two fractions have the same numerator, don't add them by adding the denominators and leaving the numerator unchanged.

Adding fractions with different denominators

When the fractions that you want to add have different denominators, adding them isn't quite as easy. At the same time, it doesn't have to be as hard as most teachers make it.

Now, I'm shimmying out onto a brittle limb here, but this needs to be said: Fractions can be added in a very simple way. It always works. It makes adding fractions only a little more difficult than multiplying them. And as you move up the math food chain into algebra, it becomes the most useful method.

So why doesn't anybody talk about it? I think it's a clear case of tradition being stronger than common sense. The traditional way to add fractions is more difficult, more time-consuming, and more likely to cause an error. But generation after generation has been taught that it's the right way to add fractions. It's a vicious cycle.

But in this book, I'm breaking with tradition. I first show you the easy way to add fractions. Then I show you a quick trick that works in a few special cases. Finally, I show you the traditional way to add fractions.

Using the easy way

TIP

At some point in your life, I bet some teacher somewhere told you these golden words of wisdom: "You can't add two fractions with different denominators." Your teacher was wrong! Here's the way to do it:

1. **Cross-multiply the two fractions and add the results together to get the numerator of the answer.**

   Suppose you want to add the fractions $\frac{1}{3}$ and $\frac{2}{5}$. To get the numerator of the answer, cross-multiply. In other words, multiply the numerator of each fraction by the denominator of the other:

   $$\frac{1}{3}+\frac{2}{5}$$
   $$1\times5=5$$
   $$2\times3=6$$

   Add the results to get the numerator of the answer:

   $$5+6=11$$

2. **Multiply the two denominators to get the denominator of the answer.**

   To get the denominator, just multiply the denominators of the two fractions:

   $$3\times5=15$$

   The denominator of the answer is 15.

3. Write your answer as a fraction.

$$\frac{1}{3}+\frac{2}{5}=\frac{11}{15}$$

As you discover in the earlier section "Finding the sum of fractions with the same denominator," when you add fractions, you sometimes need to reduce the answer you get. Here's an example:

$$\frac{5}{8}+\frac{3}{10}=\frac{(5\times10)+(3\times8)}{8\times10}=\frac{50+24}{80}=\frac{74}{80}$$

Because the numerator and the denominator are both even numbers, you know that the fraction can be reduced. So try dividing both numbers by 2:

$$\frac{74\div2}{80\div2}=\frac{37}{40}$$

This fraction can't be reduced further, so $\frac{37}{40}$ is the final answer.

As you also discover in "Finding the sum of fractions with the same denominator," sometimes when you add two proper fractions, your answer is an improper fraction:

$$\frac{4}{5}+\frac{3}{7}=\frac{(4\times7)+(3\times5)}{5\times7}=\frac{28+15}{35}=\frac{43}{35}$$

If you have more work to do with this fraction, leave it as an improper fraction so that it's easier to work with. But if this is your final answer, you may need to turn it into a mixed number (see Chapter 9 for details).

$$\frac{43}{35}=43\div35=1\text{r}8=1\frac{8}{35}$$

In some cases, you have to add more than one fraction. The method is similar, with one small tweak. For example, suppose you want to add $\frac{1}{2}+\frac{3}{5}+\frac{4}{7}$:

1. **Start by multiplying the *numerator* of the first fraction by the *denominators* of all the other fractions.**

$$\frac{\mathbf{1}}{2}+\frac{3}{\mathbf{5}}+\frac{4}{\mathbf{7}}$$
$$(1\times5\times7)=35$$

2. **Do the same with the second fraction, and add this value to the first.**

$$\frac{1}{2}+\frac{3}{5}+\frac{4}{7}$$
$$35+(3\times2\times7)=35+42$$

3. **Do the same with the remaining fraction(s).**

$$\frac{1}{2}+\frac{3}{5}+\frac{4}{7}$$
$$35+42+(4\times2\times5)=35+42+40=117$$

When you're done, you have the numerator of the answer.

4. **To get the denominator, just multiply all the denominators together:**

$$\frac{1}{2}+\frac{3}{5}+\frac{4}{7}$$
$$=\frac{35+42+40}{2\times5\times7}=\frac{117}{70}$$

As usual, you may need to reduce or change an improper fraction to a mixed number. In this example, you just need to change to a mixed number (see Chapter 9 for details):

$$\frac{117}{70}=117\div70=1\text{ r }47=1\frac{47}{70}$$

Trying a quick trick

I show you a way to add fractions with different denominators in the preceding section. It always works, and it's easy. So why do I want to show you another way? It feels like déjà vu.

In some cases, you can save yourself a lot of effort with a little bit of smart thinking. You can't always use this method, but you can use it when one denominator is a multiple of the other. (For more on multiples, see Chapter 8.) Look at the following problem:

$$\frac{11}{12}+\frac{19}{24}$$

First, I solve it the way I show you in the preceding section:

$$\frac{11}{12}+\frac{19}{24}=\frac{(11\times24)+(19\times12)}{12\times24}=\frac{264+228}{288}=\frac{492}{288}$$

Those numbers are pretty big, and I'm still not done because the numerator is larger than the denominator. The answer is an improper fraction. Worse yet, the numerator and denominator are both even numbers, so the answer still needs to be reduced.

With certain fraction addition problems, I can give you a smarter way to work. The trick is to turn a problem with different denominators into a much easier problem with the same denominator.

Before you add two fractions with different denominators, check the denominators to see whether one is a multiple of the other (for more on multiples, flip to Chapter 8). If it is, you can use the quick trick:

1. **Increase the terms of the fraction with the smaller denominator so that it has the larger denominator.**

   Look at the earlier problem in this new way:

   $$\frac{11}{12}+\frac{19}{24}$$

   As you can see, 12 divides into 24 without a remainder. In this case, you want to raise the terms of $\frac{11}{12}$ so that the denominator is 24:

   $$\frac{11}{12}=\frac{?}{24}$$

   I show you how to do this kind of problem in Chapter 9. To fill in the question mark, the trick is to divide 24 by 12 to find out how the denominators are related; then multiply the result by 11:

   $$? = (24 \div 12) \times 11 = 22$$

   $$\text{So } \frac{11}{12}=\frac{22}{24}$$

2. **Rewrite the problem, substituting this increased version of the fraction, and add as I show you earlier in "Finding the sum of fractions with the same denominator."**

   Now you can rewrite the problem this way:

   $$\frac{22}{24}+\frac{19}{24}=\frac{41}{24}$$

As you can see, the numbers in this case are much smaller and easier to work with. The answer here is an improper fraction; changing it to a mixed number is easy:

$$\frac{41}{24} = 41 \div 24 = 1\text{r}17 = 1\frac{17}{24}$$

Relying on the traditional way

In the two preceding sections, I show you two ways to add fractions with different denominators. They both work great, depending on the circumstances. So why do I want to show you yet a third way? It feels like déjà vu all over again.

The truth is that I don't want to show you this way. But they're *forcing* me to. And you know who *they* are, don't you? The man — the system — the powers that be. The ones who want to keep you down in the mud, groveling at their feet. Okay, so I'm exaggerating a little. But let me impress on you that you don't have to add fractions this way unless you really want to (or unless your teacher insists on it).

Here's the traditional way to add fractions with two different denominators:

1. **Find the least common multiple (LCM) of the two denominators (for more on finding the LCM of two numbers, see Chapter 8).**

   Suppose you want to add the fractions $\frac{3}{4} + \frac{7}{10}$. First find the LCM of the two denominators, 4 and 10. Here's how to find the LCM using the multiplication table method:

   - **Multiples of 10:** 10, 20, 30, 40
   - **Multiples of 4:** 4, 8, 12, 16, 20

   So the LCM of 4 and 10 is 20.

2. **Increase the terms of each fraction so that the denominator of each equals the LCM (for more on how to do this, see Chapter 9).**

   Increase each fraction to higher terms so that the denominator of each is 20.

   $$\frac{3}{4} = \frac{3 \times 5}{4 \times 5} = \frac{15}{20} \text{ and } \frac{7}{10} = \frac{7 \times 2}{10 \times 2} = \frac{14}{20}$$

3. **Substitute these two new fractions for the original ones and add as I show you earlier in "Finding the sum of fractions with the same denominator."**

   At this point, you have two fractions that have the same denominator:

   $$\frac{15}{20}+\frac{14}{20}=\frac{29}{20}$$

   When the answer is an improper fraction, you still need to change it to a mixed number:

   $$\frac{29}{20}=29\div 20=1\text{r}9=1\frac{9}{20}$$

As another example, suppose you want to add the numbers $\frac{5}{6}+\frac{3}{10}+\frac{2}{15}$.

1. **Find the LCM of 6, 10, and 15.**

   This time, I use the prime factorization method (see Chapter 8 for details on how to do this). Start by decomposing the three denominators to their prime factors:

   $$6=2\times 3$$
   $$10=2\times 5$$
   $$15=3\times 5$$

   These denominators have a total of three different prime factors — 2, 3, and 5. Each prime factor appears only once in any decomposition, so the LCM of 6, 10, and 15 is

   $$2\times 3\times 5=30$$

2. **You need to increase the terms of all three fractions so that their denominators are 30:**

   $$\frac{5}{6}=\frac{5\times 5}{6\times 5}=\frac{25}{30}$$
   $$\frac{3}{10}=\frac{3\times 3}{10\times 3}=\frac{9}{30}$$
   $$\frac{2}{15}=\frac{2\times 2}{15\times 2}=\frac{4}{30}$$

3. **Simply add the three new fractions:**

   $$\frac{25}{30}=\frac{9}{30}+\frac{4}{30}=\frac{38}{30}$$

Again, you need to change this improper fraction to a mixed number:

$$\frac{38}{30} = 38 \div 30 = 1 \text{ r } 8 = 1\frac{8}{30}$$

Because both numbers are divisible by 2, you can reduce the fraction:

$$1\frac{8}{30} = 1\frac{4}{15}$$

Picking your trick: Choosing the best method

As I say earlier in this chapter, I think the traditional way to add fractions is more difficult than either the easy way or the quick trick. Your teacher may require you to use the traditional way, and after you get the hang of it, you'll get good at it. But given the choice, here's my recommendation:

- Use the easy way when the numerators and denominators are small (say, 15 or under).
- Use the quick trick with larger numerators and denominators when one denominator is a multiple of the other.
- Use the traditional way only when you can't use either of the other methods (or when you know the LCM just by looking at the denominators).

Taking It Away: Subtracting Fractions

Subtracting fractions isn't really much different than adding them. As with addition, when the denominators are the same, subtraction is easy. And when the denominators are different, the methods I show you for adding fractions can be tweaked for subtracting them.

So to figure out how to subtract fractions, you can read the section "All Together Now: Adding Fractions" and substitute a minus sign (–) for every plus sign (+). But it'd be just a little chintzy if I expected you to do that. So in this section, I show you four ways to subtract fractions that mirror what I discuss earlier in this chapter about adding them.

Subtracting fractions with the same denominator

As with addition, subtracting fractions with the same denominator is always easy. When the denominators are the same, you can just think of the fractions as pieces of cake.

To subtract one fraction from another when they both have the same denominator (bottom number), subtract the numerator (top number) of the second from the numerator of the first and keep the denominator the same. For example:

$$\frac{3}{5} - \frac{2}{5} = \frac{3-2}{5} = \frac{1}{5}$$

Sometimes, as when you add fractions, you have to reduce:

$$\frac{3}{10} - \frac{1}{10} = \frac{3-1}{10} = \frac{2}{10}$$

Because the numerator and denominator are both even, you can reduce this fraction by a factor of 2:

$$\frac{2}{10} = \frac{2 \div 2}{10 \div 2} = \frac{1}{5}$$

Unlike addition, when you subtract one proper fraction from another, you never get an improper fraction.

Subtracting fractions with different denominators

Just as with addition, you have a choice of methods when subtracting fractions. These three methods are similar to the methods I show you for adding fractions: the easy way, the quick trick, and the traditional way.

The easy way always works, and I recommend this method for most of your fraction subtracting needs. The quick trick is a great timesaver, so use it when you can. And as for the traditional way — well, even if I don't like it, your teacher and other math purists probably do.

Knowing the easy way

This way of subtracting fractions works in all cases, and it's easy. (In the next section, I show you a quick way to subtract fractions when one denominator is a multiple of the other.) Here's the easy way to subtract fractions that have different denominators:

1. **Cross-multiply the two fractions and subtract the second number from the first to get the numerator of the answer.**

   For example, suppose you want to subtract $\frac{6}{7}-\frac{2}{5}$. To get the numerator, cross-multiply the two fractions and then subtract the second number from the first number (see Chapter 9 for info on cross-multiplication):

   $$\frac{6}{7}-\frac{2}{5}$$
   $$(6\times5)-(2\times7)=30-14=16$$

   After you cross-multiply, be sure to subtract in the correct order. (The first number is the numerator of the first fraction times the denominator of the second.)

2. **Multiply the two denominators to get the denominator of the answer.**

   $$7\times5=35$$

3. **Putting the numerator over the denominator gives you your answer.**

   $$\frac{16}{35}$$

Here's another example to work with:

$$\frac{9}{10}-\frac{5}{6}$$

This time, I put all the steps together:

$$\frac{9}{10}-\frac{5}{6}=\frac{(9\times6)-(5\times10)}{10\times6}$$

With the problem set up like this, you just have to simplify the result:

$$=\frac{54-50}{60}=\frac{4}{60}$$

In this case, you can reduce the fraction:

$$\frac{4}{60} = \frac{1}{15}$$

Cutting it short with a quick trick

The easy way I show you in the preceding section works best when the numerators and denominators are small. When they're larger, you may be able to take a shortcut.

Before you subtract fractions with different denominators, check the denominators to see whether one is a multiple of the other (for more on multiples, see Chapter 8). If it is, you can use the quick trick:

1. **Increase the terms of the fraction with the smaller denominator so that it has the larger denominator.**

   For example, suppose you want to find $\frac{17}{20} - \frac{31}{80}$. If you cross-multiply these fractions, your results are going to be much bigger than you want to work with. But fortunately, 80 is a multiple of 20, so you can use the quick way.

   First, increase the terms of $\frac{17}{20}$ so that the denominator is 80 (for more on increasing the terms of fractions, see Chapter 9):

   $$\frac{17}{20} = \frac{?}{80}$$

   $$? = 80 \div 20 \times 17 = 68$$

   $$\text{So } \frac{17}{20} = \frac{68}{80}$$

2. **Rewrite the problem, substituting this increased version of the fraction, and subtract as I show you earlier in "Subtracting fractions with the same denominator."**

   Here's the problem as a subtraction of fractions with the same denominator, which is much easier to solve:

   $$\frac{68}{80} - \frac{31}{80} = \frac{37}{80}$$

   In this case, you don't have to reduce to lowest terms, although you may have to in other problems. (See Chapter 9 for more on reducing fractions.)

Keeping your teacher happy with the traditional way

As I describe earlier in this chapter in "All Together Now: Adding Fractions," you want to use the traditional way only as a last resort. I recommend that you use it only when the numerator and denominator are too large to use the easy way and when you can't use the quick trick.

To use the traditional way to subtract fractions with two different denominators, follow these steps:

1. **Find the least common multiple (LCM) of the two denominators (for more on finding the LCM of two numbers, see Chapter 8).**

   For example, suppose you want to subtract $\frac{7}{8}-\frac{11}{14}$. Here's how to find the LCM of 8 and 14:

   $$\text{Multiples of } 8: \ 8, \ 16, \ 24, \ 32, \ 40, \ 48, \ 56$$
   $$\text{Multiples of } 14: \ 14, \ 28, \ 42, \ 56$$

   So the LCM of 8 and 14 is 56.

2. **Increase each fraction to higher terms so that the denominator of each equals the LCM (for more on how to do this, see Chapter 9).**

   The denominators of both now are 56:

   $$\frac{7}{8}=\frac{7\times7}{8\times7}=\frac{49}{56}$$
   $$\frac{11}{14}=\frac{11\times4}{14\times4}=\frac{44}{56}$$

3. **Substitute these two new fractions for the original ones and subtract as I show you earlier in "Subtracting fractions with the same denominator."**

   $$\frac{49}{56}-\frac{44}{56}=\frac{5}{56}$$

   This time, you don't need to reduce because 5 is a prime number and 56 isn't divisible by 5. In some cases, however, you have to reduce the answer to lowest terms.

Working Properly with Mixed Numbers

All the methods I describe earlier in this chapter work for both proper and improper fractions. Unfortunately, mixed numbers are ornery little critters, and you need to figure out how to deal with them on their own terms. (For more on mixed numbers, flip to Chapter 9.)

Multiplying and dividing mixed numbers

I can't give you a direct method for multiplying and dividing mixed numbers. The only way is to convert the mixed numbers to improper fractions and multiply or divide as usual. Here's how to multiply or divide mixed numbers:

1. **Convert all mixed numbers to improper fractions (see Chapter 9 for details).**

   For example, suppose you want to multiply $1\frac{1}{5} \times 2\frac{1}{3}$. First convert $1\frac{3}{5}$ and $2\frac{1}{3}$ to improper fractions:

   $$1\frac{3}{5} = \frac{5 \times 1 + 3}{5} = \frac{8}{5}$$
   $$2\frac{1}{3} = \frac{3 \times 2 + 1}{3} = \frac{7}{3}$$

2. **Multiply these improper fractions (as I show you earlier in this chapter, in "Multiplying and Dividing Fractions").**

   $$\frac{8}{5} \times \frac{7}{3} = \frac{8 \times 7}{5 \times 3} = \frac{56}{15}$$

3. **If the answer is an improper fraction, convert it back to a mixed number (see Chapter 9).**

   $$\frac{56}{15} = 56 \div 15 = 3\text{r}11 = 3\frac{11}{15}$$

   In this case, the answer is already in lowest terms, so you don't have to reduce it.

As a second example, suppose you want to divide $3\frac{2}{3}$ by $1\frac{4}{7}$.

1. **Convert $3\frac{2}{3}$ and $1\frac{4}{7}$ to improper fractions:**

   $$3\frac{2}{3} = \frac{3 \times 3 + 2}{3} = \frac{11}{3}$$
   $$1\frac{4}{7} = \frac{7 \times 1 + 4}{7} = \frac{11}{7}$$

2. **Divide these improper fractions.**

   Divide fractions by multiplying the first fraction by the reciprocal of the second (see the earlier "Multiplying and Dividing Fractions" section):

   $$\frac{11}{3} \div \frac{11}{7} = \frac{11}{3} \times \frac{7}{11}$$

   In this case, before you multiply, you can cancel out factors of 11 in the numerator and denominator:

   $$\frac{1\not{11}}{3} \times \frac{7}{\not{11}1} = \frac{1 \times 7}{3 \times 1} = \frac{7}{3}$$

3. **Convert the answer to a mixed number.**

   $$\frac{7}{3} = 7 \div 3 = 2\text{r}1 = 2\frac{1}{3}$$

Adding and subtracting mixed numbers

One way to add and subtract mixed numbers is to convert them to improper fractions, much as I describe earlier in this chapter in "Multiplying and dividing mixed numbers," and then to add or subtract them using a method from the "All Together Now: Adding Fractions" or "Take It Away: Subtracting Fractions" sections. Doing so is a perfectly valid way of getting the right answer without learning a new method.

Unfortunately, teachers just love to make people add and subtract mixed numbers in their own special way. The good news is that a lot of folks find this way easier than all the converting stuff.

Working in pairs: Adding two mixed numbers

Adding mixed numbers looks a lot like adding whole numbers: You stack them one on top of the other, draw a line, and add. For this reason, some students feel more comfortable adding mixed numbers than adding fractions. Here's how to add two mixed numbers:

1. **Add the fractional parts using any method you like; if necessary, change this sum to a mixed number and reduce it.**
2. **If the answer you found in Step 1 is an improper fraction, change it to a mixed number, write down the fractional part, and carry the whole number part to the whole number column.**

3. Add the whole number parts (including any number carried).

You may also need to reduce your answer to lowest terms (see Chapter 9). In the examples that follow, I show you everything you need to know.

Summing up mixed numbers when the denominators are the same

As with any problem involving fractions, adding is always easier when the denominators are the same. For example, suppose you want to add $3\frac{1}{3}+5\frac{1}{3}$. Doing mixed number problems is often easier if you place one number above the other:

$$\begin{array}{r} 3\frac{1}{3} \\ +5\frac{1}{3} \\ \hline \end{array}$$

As you can see, this arrangement is similar to how you add whole numbers, but it includes an extra column for fractions. Here's how you add these two mixed numbers step by step:

1. **Add the fractions.**

   $$\frac{1}{3}+\frac{1}{3}=\frac{2}{3}$$

2. **Switch improper fractions to mixed numbers; write down your answer.**

   Because $\frac{2}{3}$ is a proper fraction, you don't have to change it.

3. **Add the whole number parts.**

   $$3+5=8$$

Here's how your problem looks in column form:

$$\begin{array}{r} 3\frac{1}{3} \\ +5\frac{1}{3} \\ \hline 8\frac{2}{3} \end{array}$$

This problem is about as simple as they get. In this case, all three steps are pretty easy. But sometimes, Step 2 requires more attention. For example, suppose you want to add $8\frac{3}{5}+6\frac{4}{5}$. Here's how you do it:

1. **Add the fractions.**

   $$\frac{3}{5}+\frac{4}{5}=\frac{7}{5}$$

2. **Switch improper fractions to mixed numbers, write down the fractional part, and carry over the whole number.**

   Because the sum is an improper fraction, convert it to the mixed number $1\frac{2}{5}$ (flip to Chapter 9 for more on converting improper fractions to mixed numbers). Write down $\frac{2}{5}$ and carry the 1 over to the whole number column.

3. **Add the whole number parts, including any whole numbers you carried over when you switched to a mixed number.**

   $$1 + 8 + 6 = 15$$

Here's how the solved problem looks in column form. (Be sure to line up the whole numbers in one column and the fractions in another.)

$$\begin{array}{r} 1 \\ 8\frac{3}{5} \\ +6\frac{4}{5} \\ \hline 15\frac{2}{5} \end{array}$$

As with any other problems involving fractions, sometimes you need to reduce at the end of Step 1.

The same basic idea works no matter how many mixed numbers you want to add. For example, suppose you want to add $5\frac{4}{9} + 11\frac{7}{9} + 3\frac{8}{9} + 1\frac{5}{9}$:

1. **Add the fractions.**

   $$\frac{4}{9} + \frac{7}{9} + \frac{8}{9} + \frac{5}{9} = \frac{24}{9}$$

2. **Switch improper fractions to mixed numbers, write down the fractional part, and carry over the whole number.**

   Because the result is an improper fraction, convert it to the mixed number $2\frac{6}{9}$ and then reduce it to $2\frac{2}{3}$ (for more on converting and reducing fractions, see Chapter 9). I recommend doing these calculations on a piece of scrap paper.

   Write down $\frac{2}{3}$ and carry the 2 to the whole number column.

3. **Add the whole numbers.**

   $$2 + 5 + 11 + 3 + 1 = 22$$

Here's how the problem looks after you solve it:

$$\begin{array}{r} 2\phantom{\frac{0}{0}} \\ 5\frac{4}{9} \\ 11\frac{7}{9} \\ 3\frac{8}{9} \\ +1\frac{5}{9} \\ \hline 22\frac{2}{3} \end{array}$$

Summing up mixed numbers when the denominators are different

The most difficult type of mixed number addition is when the denominators of the fractions are different. This difference doesn't change Steps 2 or 3, but it does make Step 1 tougher.

For example, suppose you want to add $16\frac{3}{5}$ and $7\frac{7}{9}$.

1. **Add the fractions.**

   Add $\frac{3}{5}$ and $\frac{7}{9}$. You can use any method from earlier in this chapter. Here, I use the easy way:

   $$\frac{3}{5}+\frac{7}{9}=\frac{(3\times9)+(7\times5)}{5\times9}=\frac{27+35}{45}=\frac{62}{45}$$

2. **Switch improper fractions to mixed numbers, write down the fractional part, and carry over the whole number.**

   This fraction is improper, so change it to the mixed number $1\frac{17}{45}$. Fortunately, the fractional part of this mixed number isn't reducible. Write down the $\frac{17}{45}$ and carry over the 1 to the whole number column.

3. **Add the whole numbers.**

   $$1+16+7=24$$

Here's how the completed problem looks:

$$\begin{array}{r} 1\phantom{\frac{0}{0}} \\ 16\frac{3}{5} \\ +7\frac{7}{9} \\ \hline 24\frac{17}{45} \end{array}$$

Subtracting mixed numbers

The basic way to subtract mixed numbers is close to the way you add them. Again, the subtraction looks more like what you're used to with whole numbers. Here's how to subtract two mixed numbers:

1. **Find the difference of the fractional parts using any method you like.**
2. **Find the difference of the two whole number parts.**

Along the way, though, you may encounter a couple more twists and turns. I keep you on track so that, by the end of this section, you can do any mixed-number subtraction problem.

Taking away mixed numbers when the denominators are the same

As with addition, subtraction is much easier when the denominators are the same. For example, suppose you want to subtract $7\frac{3}{5} - 3\frac{1}{5}$. Here's what the problem looks like in column form:

$$\begin{array}{r} 7\frac{3}{5} \\ -3\frac{1}{5} \\ \hline 4\frac{2}{5} \end{array}$$

In this problem, I subtract $\frac{3}{5} - \frac{1}{5} = \frac{2}{5}$. Then I subtract $7 - 3 = 4$. Not too ter rible, agreed?

One complication arises when you try to subtract a larger fractional part from a smaller one. Suppose you want to find $11\frac{1}{6} - 2\frac{5}{6}$. This time, if you try to subtract the fractions, you get

$$\frac{1}{6} - \frac{5}{6} = -\frac{4}{6}$$

Obviously, you don't want to end up with a negative number in your answer. You can handle this problem by borrowing from the column to the left. This idea is similar to the borrowing that you use in regular subtraction, with one key difference.

When borrowing in mixed-number subtraction,

1. **Borrow 1 from the whole-number portion and add it to the fractional portion, turning the fraction into a mixed number.**

   To find $11\frac{1}{6} - 2\frac{5}{6}$, borrow 1 from the 11 and add it to $\frac{1}{6}$, making it the mixed number $1\frac{1}{6}$:

$$11\frac{1}{6} = 10 + 1\frac{1}{6}$$

2. **Change this new mixed number into an improper fraction.**

   Here's what you get when you change $1\frac{1}{6}$ into an improper fraction:

   $$10 + 1\frac{1}{6} = 10\frac{7}{6}$$

   The result is $10\frac{7}{6}$. This answer is a weird cross between a mixed number and an improper fraction, but it's what you need to handle the job.

3. **Use the result in your subtraction.**

   $$\begin{array}{r} 10\frac{7}{6} \\ -2\frac{5}{6} \\ \hline 8\frac{2}{6} \end{array}$$

   In this case, you have to reduce the fractional part of the answer:

   $$8\frac{2}{6} = 8\frac{1}{3}$$

Subtracting mixed numbers when the denominators are different

Subtracting mixed numbers when the denominators are different is just about the hairiest thing you're ever going to have to do in pre-algebra. Fortunately, though, if you work through this chapter, you acquire all the skills you need.

Suppose you want to subtract $15\frac{4}{11} - 12\frac{3}{7}$. Because the denominators are different, subtracting the fractions becomes more difficult. But you have another question to think about: In this problem, do you need to borrow? If $\frac{4}{11}$ is greater than $\frac{3}{7}$, you don't have to borrow. But if $\frac{4}{11}$ is less than $\frac{3}{7}$, you do. (For more on borrowing in mixed-number subtraction, see the preceding section.)

In Chapter 9, I show you how to test two fractions to see which is greater by cross-multiplying:

$$4 \times 7 = 28$$
$$3 \times 11 = 33$$

Because 28 is less than 33, $\frac{4}{11}$ is less than $\frac{3}{7}$, so you do have to borrow. I get the borrowing out of the way first:

$$15\frac{4}{11} = 14 + 1\frac{4}{11} = 14\frac{15}{11}$$

Now the problem looks like this:

$$14\frac{15}{11} - 12\frac{3}{7}$$

The first step, subtracting the fractions, is the most time-consuming, so as I show you earlier in "Subtracting fractions with different denominators," you can take care of that on the side:

$$\frac{15}{11} - \frac{3}{7} = \frac{(15 \times 7) - (3 \times 11)}{11 \times 7} = \frac{105 - 33}{77} = \frac{72}{77}$$

The good news is that this fraction can't be reduced (72 and 77 have no common factors: $72 = 2 \times 2 \times 2 \times 3 \times 3$ and $77 = 7 \times 11$). So the hard part of the problem is done, and the rest follows easily:

$$\begin{array}{r} 14\frac{15}{11} \\ -12\ \frac{3}{7} \\ \hline 2\frac{72}{77} \end{array}$$

This problem is about as difficult as a mixed-number subtraction problem gets. Take a look at it step by step. Better yet, copy the problem and then close the book and try to work through the steps on your own. If you get stuck, that's okay: Better now than on an exam!

Chapter 11

Dallying with Decimals

In This Chapter

- Understanding the decimal basics
- Applying decimals to the Big Four operations
- Looking at decimal and fraction conversions
- Making sense of repeating decimals

Because early humans used their fingers for counting, the number system is based on the number 10. So numbers come in ones, tens, hundreds, thousands, and so on. A *decimal* — with its handy decimal point — allows people to work with numbers smaller than 1: tenths, hundredths, thousandths, and the like.

Here's some lovely news: Decimals are much easier to work with than fractions (which I discuss in Chapters 9 and 10). Decimals look and feel more like whole numbers than fractions do, so when you're working with decimals, you don't have to worry about reducing and increasing terms, improper fractions, mixed numbers, and a lot of other stuff.

Performing the Big Four operations — addition, subtraction, multiplication, and division — on decimals is very close to performing them on whole numbers (which I cover in Part II of the book). The numerals 0 through 9 work just like they usually do. As long as you get the decimal point in the right place, you're home free.

In this chapter, I show you all about working with decimals. I also show you how to convert fractions to decimals and decimals to fractions. Finally, I give you a peek into the strange world of repeating decimals.

Understanding Basic Decimal Stuff

The good news about decimals is that they look a lot more like whole numbers than fractions do. So a lot of what you find out about whole numbers in Chapter 2 applies to decimals as well. In this section, I introduce you to decimals, starting with place value.

When you understand place value of decimals, a lot falls into place. Then I discuss trailing zeros and what happens when you move the decimal point either to the left or to the right.

Counting dollars and decimals

You use decimals all the time when you count money. And a great way to begin thinking about decimals is as dollars and cents. For example, you know that $0.50 is half of a dollar (see Figure 11-1), so this information tells you:

$$0.5 = \frac{1}{2}$$

Figure 11-1: One-half (0.5) of a dollar bill.

Illustration by Wiley, Composition Services Graphics

Notice that, in the decimal 0.5, I drop the zero at the end. This practice is common with decimals.

You also know that $0.25 is a quarter — that is, one-fourth of a dollar (see Figure 11-2) — so:

$$0.25 = \frac{1}{4}$$

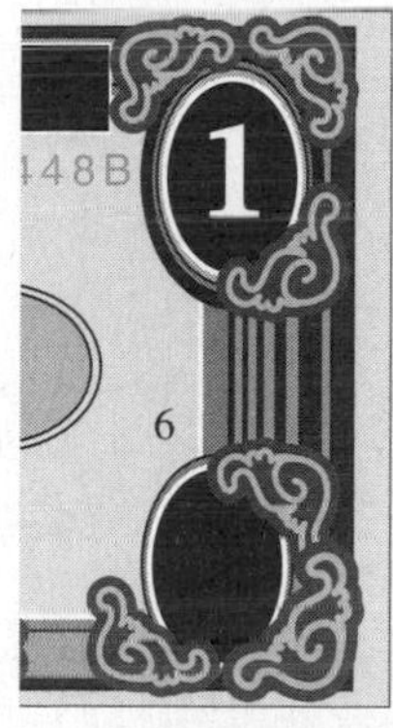

Figure 11-2: One-fourth (0.25) of a dollar bill.

Illustration by Wiley, Composition Services Graphics

Similarly, you know that $0.75 is three quarters, or three-fourths, of a dollar (see Figure 11-3), so:

$$0.75 = \frac{3}{4}$$

Figure 11-3: Three-fourths (0.75) of a dollar bill.

Illustration by Wiley, Composition Services Graphics

Taking this idea even further, you can use the remaining denominations of coins — dimes, nickels, and pennies — to make further connections between decimals and fractions.

$$\text{A dime} = \$0.10 = \frac{1}{10} \text{ of a dollar, so } \frac{1}{10} = 0.1$$

$$\text{A nickel} = \$0.05 = \frac{1}{20} \text{ of a dollar, so } \frac{1}{20} = 0.05$$

$$\text{A penny} = \$0.01 = \frac{1}{100} \text{ of a dollar, so } \frac{1}{100} = 0.01$$

Notice that I again drop the final zero in the decimal 0.1, but I keep the zeros in the decimals 0.05 and 0.01. You can drop zeros from the right end of a decimal, but you can't drop zeros that fall between the decimal point and another digit.

Decimals are just as good for cutting up cake as for cutting up money. Figure 11-4 gives you a look at the four cut-up cakes that I show you in Chapter 9. This time, I give you the decimals that tell you how much cake you have. Fractions and decimals accomplish the same task: allowing you to cut a whole object into pieces and talk about how much you have.

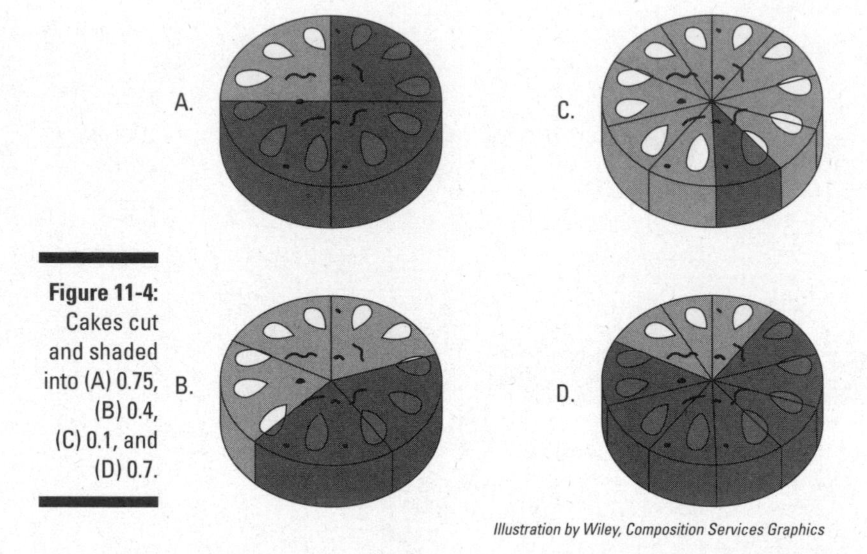

Figure 11-4: Cakes cut and shaded into (A) 0.75, (B) 0.4, (C) 0.1, and (D) 0.7.

Illustration by Wiley, Composition Services Graphics

Identifying the place value of decimals

In Chapter 2, you find out about the place value of whole numbers. Table 11-1 shows how the whole number 4,672 breaks down in terms of place value.

Table 11-1 Breaking Down 4,672 in Terms of Place Value

Thousands	*Hundreds*	*Tens*	*Ones*
4	6	7	2

This number means 4,000 + 600 + 70 + 2.

With decimals, this idea is extended. First, a decimal point is placed to the right of the ones place in a whole number. Then more numbers are appended to the right of the decimal point.

For example, the decimal 4,672.389 breaks down as shown in Table 11-2.

Table 11-2 Breaking Down the Decimal 4,672.389

Thousands	*Hundreds*	*Tens*	*Ones*	*Decimal Point*	*Tenths*	*Hundredths*	*Thousandths*
4	6	7	2	.	3	8	9

This decimal means $4{,}000+600+70+2+\frac{3}{10}+\frac{8}{100}+\frac{9}{1000}$.

The connection between fractions and decimals becomes obvious when you look at place value. Decimals really are a shorthand notation for fractions. You can represent any fraction as a decimal.

Knowing the decimal facts of life

When you understand how place value works in decimals (as I explain in the preceding section), a whole lot of facts about decimals begin to make sense. Two key ideas are trailing zeros and what happens when you move a decimal point left or right.

Understanding trailing zeros

You probably know that you can attach zeros to the beginning of a whole number without changing its value. For example, these three numbers are all equal in value:

27 027 0,000,027

The reason becomes clear when you know about place value of whole numbers. See Table 11-3.

Table 11-3 Example of Attaching Leading Zeros

Millions	*Hundred Thousands*	*Ten Thousands*	*Thousands*	*Hundreds*	*Tens*	*Ones*
0	0	0	0	0	2	7

As you can see, 0,000,027 simply means 0 + 0 + 0 + 0 + 0 + 20 + 7. No matter how many zeros you add to the beginning of a number, the number 27 doesn't change.

Zeros attached to the beginning of a number in this way are called *leading zeros.*

In decimals, this idea of zeros that don't add value to a number can be extended to trailing zeros.

A *trailing zero* is any zero that appears to the right of both the decimal point and every digit other than zero.

For example:

34.8 34.80 34.8000

All three of these numbers are the same. The reason becomes clear when you understand how place value works in decimals. See Table 11-4.

Table 11-4 Example of Attaching Trailing Zeros

Tens	*Ones*	*Decimal Point*	*Tenths*	*Hundredths*	*Thousandths*	*Ten Thousandths*
3	4	.	8	0	0	0

In this example, 34.8000 means $40+4+\frac{8}{10}+\frac{0}{100}+\frac{0}{1000}+\frac{0}{10000}$.

You can attach or remove as many trailing zeros as you want without changing the value of a number.

When you understand trailing zeros, you can see that every whole number can easily be changed to a decimal. Just attach a decimal point and a 0 to the end of it. For example:

$$4 = 4.0$$
$$20 = 20.0$$
$$971 = 971.0$$

Make sure that you don't attach or remove any nonleading or nontrailing zeros — it changes the value of the decimal.

For example, look at this number:

0450.0070

In this number, you can remove the leading and trailing zeros without changing the value, as follows:

450.007

The remaining zeros, however, need to stay where they are as *placeholders* between the decimal point and digits other than zero. See Table 11-5.

Table 11-5 Example of Zeros as Placeholders

Thousands	*Hundreds*	*Tens*	*Ones*	*Decimal Point*	*Tenths*	*Hundredths*	*Thousandths*	*Ten Thousandths*
0	4	5	0	.	0	0	7	0

I continue to discuss zeros as placeholders in the next section.

Moving the decimal point

When you're working with whole numbers, you can multiply any number by 10 just by adding a zero to the end of it. For example:

$$45,971 \times 10 = 459,710$$

To see why this answer is so, again think about the place value of digits and look at Table 11-6.

Table 11-6 Example, Decimal Points and Place Value of Digits

Millions	*Hundred Thousands*	*Ten Thousands*	*Thousands*	*Hundreds*	*Tens*	*Ones*
		4	5	9	7	1
	4	5	9	7	1	0

Here's what these two numbers really mean:

$$45,971 = 40,000 + 5,000 + 900 + 70 + 1$$
$$459,710 = 400,000 + 50,000 + 9,000 + 700 + 10 + 0$$

As you can see, that little zero makes a big difference: It causes the rest of the numbers to shift one place.

This concept makes even more sense when you think about the decimal point. See Table 11-7.

Table 11-7 Example, Numbers Shifting One Place

Hundred Thousands	*Ten Thousands*	*Thousands*	*Hundreds*	*Tens*	*Ones*	*Decimal Point*	*Tenths*	*Hundredths*
	4	5	9	7	1	.	0	0
4	5	9	7	1	0	.	0	0

In effect, adding a 0 to the end of a whole number moves the decimal point one place to the right. So for any decimal, when you move the decimal point one place to the right, you multiply that number by 10. This fact becomes clear when you start with a simple number like 7:

7.0

70.0

700.0

7,000.0

In this case, the net effect is that you moved the decimal point three places to the right, which is the same as multiplying 7 by 1,000.

Similarly, to divide any number by 10, move the decimal point one place to the left. For example:

7.0

0.7

0.07

0.007

This time, the net effect is that you moved the decimal point three places to the left, which is the same as dividing 7 by 1,000.

Rounding decimals

Rounding decimals works almost exactly the same as rounding numbers. You'll use this skill when dividing decimals later in the chapter. Most commonly, you need to round a decimal either to a whole number or to one or two decimal places.

To round a decimal to a whole number, focus on the ones digit and the tenths digit. Round the decimal either up or down to the *nearest* whole number, dropping the decimal point:

$$\underline{7.1} \rightarrow 7 \quad 3\underline{2.9} \rightarrow 33 \quad 18\underline{4.3} \rightarrow 184$$

When the tenths digit is 5, round the decimal *up*:

$$8\underline{3.5} \rightarrow 84 \quad 29\underline{6.5} \rightarrow 297 \quad 1{,}78\underline{8.5} \rightarrow 1{,}789$$

If the decimal has other decimal digits, just drop them:

$$1\underline{8.4}7 \rightarrow 18 \quad 2\underline{1.6}18 \rightarrow 22 \quad \underline{3.1}415927 \rightarrow 3$$

Occasionally, a small change to the ones digit affects the other digits. (This example may remind you of when the odometer in your car rolls a bunch of 9s over to 0s):

$$9\underline{9.9} \rightarrow 100 \quad 99\underline{9.5} \rightarrow 1{,}000 \quad 99{,}99\underline{9.7}12 \rightarrow 100{,}000$$

The same basic idea applies to rounding a decimal to any number of places. For example, to round a decimal to one decimal place, focus on the first and second decimal places (that is, the tenths and hundredths places):

$$76.\underline{54}3 \rightarrow 76.5 \quad 100.\underline{68}22 \rightarrow 100.7 \quad 10.\underline{10}101 \rightarrow 10.1$$

To round a decimal to two decimal places, focus on the second and third decimal places (that is, the hundredths and thousandths places):

$$444.4\underline{44}4 \rightarrow 444.44 \quad 26.5\underline{55}55 \rightarrow 26.56 \quad 99.9\underline{97} \rightarrow 100.00$$

Performing the Big Four with Decimals

Everything you already know about adding, subtracting, multiplying, and dividing whole numbers (see Chapter 3) carries over when you work with decimals. In fact, in each case, there's really only one key difference: how to handle that pesky little decimal point. In this section, I show you how to perform the Big Four math operations with decimals.

The most common use of adding and subtracting decimals is working with money — for example, balancing your checkbook. Later in this book, you find that multiplying and dividing by decimals is useful for calculating percentages (see Chapter 12), using scientific notation (see Chapter 14), and measuring with the metric system (see Chapter 15).

Adding decimals

Adding decimals is almost as easy as adding whole numbers. As long as you set up the problem correctly, you're in good shape. To add decimals, follow these steps:

1. **Arrange the numbers in a column and line up the decimal points vertically.**
2. **Add as usual, column by column, from right to left.**
3. **Place the decimal point in the answer in line with the other decimal points in the problem.**

For example, suppose you want to add the numbers 14.5 and 1.89. Line up the decimal points neatly, as follows:

$$\begin{array}{r} 14.5 \\ +\ 1.89 \\ \hline \end{array}$$

Begin adding from the right-most column. Treat the blank space after 14.5 as a 0 — you can write this in as a trailing 0 (see earlier in this chapter to see why adding zeros to the end of a decimal doesn't change its value). Adding this column gives you 0 + 9 = 9:

$$\begin{array}{r} 14.50 \\ +\ 1.89 \\ \hline 9 \end{array}$$

Continuing to the left, $5 + 8 = 13$, so put down the 3 and carry the 1:

$$\begin{array}{r} 1 \\ 14.50 \\ +\ 1.89 \\ \hline 39 \end{array}$$

Complete the problem column by column, and at the end, put the decimal point directly below the others in the problem:

$$\begin{array}{r} 14.50 \\ +\ 1.89 \\ \hline 16.39 \end{array}$$

When adding more than one decimal, the same rules apply. For example, suppose you want to add $15.1 + 0.005 + 800 + 1.2345$. The most important idea is lining up the decimal points correctly:

$$\begin{array}{r} 15.1 \\ 0.005 \\ 800.0 \\ +\ 1.2345 \\ \hline \end{array}$$

To avoid mistakes, be especially neat when adding a lot of decimals.

Because the number 800 isn't a decimal, I place a decimal point and a 0 at the end of it, to be clear about how to line it up. If you like, you can make sure all numbers have the same number of decimal places (in this case, four) by adding trailing zeros. When you properly set up the problem, the addition is no more difficult than in any other addition problem:

$$\begin{array}{r} 15.1000 \\ 0.0050 \\ 800.0000 \\ +1.2345 \\ \hline 816.3395 \end{array}$$

Subtracting decimals

Subtracting decimals uses the same trick as adding them (which I talk about in the preceding section). Here's how you subtract decimals:

1. **Arrange the numbers in a column and line up the decimal points.**
2. **Subtract as usual, column by column from right to left.**
3. **When you're done, place the decimal point in the answer in line with the other decimal points in the problem.**

For example, suppose you want to figure out 144.87 – 0.321. First, line up the decimal points:

$$\begin{array}{r} 144.870 \\ -0.321 \\ \hline \end{array}$$

In this case, I add a zero at the end of the first decimal. This placeholder reminds you that, in the right-most column, you need to borrow to get the answer to 0 – 1:

$$\begin{array}{r} \mathbf{6} \\ 144.8\not{7}\;\mathbf{10} \\ -\;0.32\;\;\;1 \\ \hline 4\;\;\;9 \end{array}$$

The rest of the problem is straightforward. Just finish the subtraction and drop the decimal point straight down:

$$\begin{array}{r} 6 \\ 144.8\not{7}\;10 \\ -\;0.32\;\;\;1 \\ \hline 144.54\;\;\;9 \end{array}$$

As with addition, the decimal point in the answer goes directly below where it appears in the problem.

Multiplying decimals

Multiplying decimals is different from adding and subtracting them, in that you don't have to worry about lining up the decimal points (see the preceding sections). In fact, the only difference between multiplying whole numbers and decimals comes at the end.

Here's how to multiply decimals:

1. **Perform the multiplication as you do for whole numbers.**
2. **When you're done, count the number of digits to the right of the decimal point in each factor, and add the result.**
3. **Place the decimal point in your answer so that your answer has the same number of digits after the decimal point.**

This process sounds tricky, but multiplying decimals can actually be simpler than adding or subtracting them. Suppose, for instance, that you want to multiply 23.5 by 0.16. The first step is to pretend that you're multiplying numbers without decimal points:

$$\begin{array}{r} 23.5 \\ \underline{\times 0.16} \\ 1410 \\ \underline{2350} \\ 3760 \end{array}$$

This answer isn't complete, though, because you still need to find out where the decimal point goes in the answer. To do this, notice that 23.5 has one digit after the decimal point and that 0.16 has two digits after the decimal point. Because 1 + 2 = 3, place the decimal point in the answer so that it has three digits after the decimal point. (You can put your pencil at the 0 at the end of 3760 and move the decimal point three places to the left.)

$$\begin{array}{rl} 23.5 & \text{1 digit after the decimal point} \\ \underline{\times 0.16} & \text{2 digits after the decimal point} \\ 1410 & \\ \underline{2350} & \\ 3760 & 1+2=3 \text{ digits after the decimal point} \end{array}$$

Even though the last digit in the answer is a 0, you still need to count this as a digit when placing the decimal point. When the decimal point is in place, you can drop trailing zeros (flip to "Understanding Basic Decimal Stuff," earlier in this chapter, to see why the zeros at the end of a decimal don't change the value of the number).

So the answer is 3.760, which is equal to 3.76.

Dividing decimals

Long division has never been a crowd pleaser. Dividing decimals is almost the same as dividing whole numbers, which is why a lot of people don't particularly like dividing decimals, either.

But at least you can take comfort in the fact that, when you know how to do long division (which I cover in Chapter 3), figuring out how to divide decimals is easy. The main difference comes at the beginning, before you start dividing.

Here's how to divide decimals:

1. **Turn the *divisor* (the number you're dividing by) into a whole number by moving the decimal point all the way to the right; at the same time, move the decimal point in the *dividend* (the number you're dividing) the same number of places to the right.**

   For example, suppose you want to divide 10.274 by 0.11. Write the problem as usual:

   $$0.11\overline{)10.274}$$

   Turn 0.11 into a whole number by moving the decimal point in 0.11 two places to the right, giving you 11. At the same time, move the decimal point in 10.274 two places to the right, giving you 1,027.4:

   $$11.\overline{)1027.4}$$

2. **Place a decimal point in the *quotient* (the answer) directly above where the decimal point now appears in the dividend.**

   Here's what this step looks like:

   $$11.\overline{)1027.4}$$

3. Divide as usual, being careful to line up the quotient properly so that the decimal point falls into place.

To start out, notice that 11 is too large to go into either 1 or 10. However, 11 does go into 102 (nine times). So write the first digit of the quotient just above the 2 and continue:

$$
\begin{array}{r}
9. \\
11.\overline{)1027.4} \\
\underline{99} \\
37
\end{array}
$$

I paused after bringing down the next number, 7. This time, 11 goes into 37 three times. The important point is to place the next digit in the answer just above the 7:

$$
\begin{array}{r}
93. \\
11.\overline{)1027.4} \\
\underline{99} \\
37 \\
\underline{33} \\
44
\end{array}
$$

I paused after bringing down the next number, 4. Now, 11 goes into 44 four times. Again, be careful to place the next digit in the quotient just above the 4, and complete the division:

$$
\begin{array}{r}
93.4 \\
11.\overline{)1027.4} \\
\underline{99} \\
37 \\
\underline{33} \\
44 \\
\underline{44} \\
0
\end{array}
$$

So the answer is 93.4. As you can see, as long as you're careful when placing the decimal point and the digits, the correct answer appears with the decimal point in the right position.

Dealing with more zeros in the dividend

Sometimes you have to add one or more trailing zeros to the dividend. As I discuss earlier in this chapter, you can add as many trailing zeros as you like to a decimal without changing its value. For example, suppose you want to divide 67.8 by 0.333:

$$0.333\overline{)67.8}$$

Follow these steps:

1. **Change 0.333 into a whole number by moving the decimal point three places to the right; at the same time, move the decimal point in 67.8 three places to the right:**

   $$333.\overline{)67800.}$$

   In this case, when you move the decimal point in 67.8, you run out of room, so you have to add a couple zeros to the dividend. This step is perfectly valid, and you need to do this whenever the divisor has more decimal places than the dividend.

2. **Place the decimal point in the quotient directly above where it appears in the dividend:**

   $$\begin{array}{r} . \\ 333.\overline{)67800.} \end{array}$$

3. **Divide as usual, being careful to correctly line up the numbers in the quotient. This time, 333 doesn't go into 6 or 67, but it does go into 678 (two times). So place the first digit of the quotient directly above the 8:**

   $$\begin{array}{r} 2. \\ 333.\overline{)67800.} \\ \underline{666} \\ 120 \end{array}$$

   I've jumped forward in the division to the place where I bring down the first 0. At this point, 333 doesn't go into 120, so you need to put a 0 above the first 0 in 67,800 and bring down the second 0. Now, 333 does go into 1,200, so place the next digit in the answer (3) over the second 0:

```
       203.
333.)67800.
     666
      1200
       999
       201
```

This time, the division doesn't work out evenly. If this were a problem with whole numbers, you'd finish by writing down a remainder of 201. (For more on remainders in division, see Chapter 3.) But decimals are a different story. The next section explains why, with decimals, the show must go on.

Completing decimal division

When you're dividing whole numbers, you can complete the problem simply by writing down the remainder. But remainders are *never* allowed in decimal division.

A common way to complete a problem in decimal division is to round off the answer. In most cases, you're instructed to round your answer to the nearest whole number or to one or two decimal places (see earlier in this chapter to find out how to round off decimals).

To complete a decimal division problem by rounding it off, you need to add at least one trailing zero to the dividend:

- To round a decimal to a whole number, add one trailing zero.
- To round a decimal to one decimal place, add two trailing zeros.
- To round a decimal to two decimal places, add three trailing zeros.

Here's what the problem looks like with a trailing zero attached:

```
       203.
333.)67800.0
     666
      1200
       999
       2010
```

Attaching a trailing zero doesn't change a decimal, but it does allow you to bring down one more number, changing 201 into 2,010. Now you can divide 333 into 2,010:

```
       203.6
333.)67800.0
     666
       1200
       999
       2010
       1998
         12
```

At this point, you can round the answer to the nearest whole number, 204. I give you more practice dividing decimals later in this chapter.

Converting between Decimals and Fractions

Fractions (see Chapters 9 and 10) and decimals are similar, in that they both allow you to represent parts of the whole — that is, these numbers fall on the number line *between* whole numbers.

In practice, though, sometimes one of these options is more desirable than the other. For example, calculators love decimals but aren't so crazy about fractions. To use your calculator, you may have to change fractions into decimals.

As another example, some units of measurement (such as inches) use fractions, whereas others (such as meters) use decimals. To change units, you may need to convert between fractions and decimals.

In this section, I show you how to convert back and forth between fractions and decimals. (If you need a refresher on fractions, review Chapters 9 and 10 before proceeding.)

Making simple conversions

Some decimals are so common that you can memorize how to represent them as fractions. Here's how to convert all the one-place decimals to fractions:

.1	.2	.3	.4	.5	.6	.7	.8	.9
$\frac{1}{10}$	$\frac{1}{5}$	$\frac{3}{10}$	$\frac{2}{5}$	$\frac{1}{2}$	$\frac{3}{5}$	$\frac{7}{10}$	$\frac{4}{5}$	$\frac{9}{10}$

And, here are few more common decimals that translate easily to fractions:

.125	.25	.375	.625	.75	.875
$\frac{1}{8}$	$\frac{1}{4}$	$\frac{3}{8}$	$\frac{5}{8}$	$\frac{3}{4}$	$\frac{7}{8}$

Changing decimals to fractions

Converting a decimal to a fraction is pretty simple. The only tricky part comes in when you have to reduce the fraction or change it to a mixed number.

In this section, I first show you the easy case, when no further work is necessary. Then I show you the harder case, when you need to tweak the fraction. I also show you a great time-saving trick.

Doing a basic decimal-to-fraction conversion

Here's how to convert a decimal to a fraction:

1. **Draw a line (fraction bar) under the decimal and place a 1 underneath it.**

   Suppose you want to turn the decimal 0.3763 into a fraction. Draw a line under 0.3763 and place a 1 underneath it:

   $$\frac{0.3763}{1}$$

   This number looks like a fraction, but technically it isn't one because the top number (the numerator) is a decimal.

2. **Move the decimal point one place to the right and add a 0 after the 1.**

   $$=\frac{3.763}{10}$$

3. **Repeat Step 2 until the decimal point moves all the way to the right so you can drop the decimal point entirely.**

   In this case, this is a three-step process:

   $$\frac{37.63}{100}=\frac{376.3}{1000}=\frac{3763}{10000}$$

As you can see on the last step, the decimal point in the numerator moves all the way to the end of the number, so dropping the decimal point is okay.

Note: Moving a decimal point one place to the right is the same thing as multiplying a number by 10. When you move the decimal point four places in this problem, you're essentially multiplying the 0.3763 and the 1 by 10,000. Notice that the number of digits after the decimal point in the original decimal is equal to the number of 0s that end up following the 1.

In the following sections, I show you how to convert decimals to fractions when you have to work with mixed numbers and reduce the terms.

Getting mixed results

When you convert a decimal greater than 1 to a fraction, the result is a mixed number. Fortunately, this process is easy because the whole number part is unaffected by the conversion. So focusing only on the decimal part, follow the same steps I outline in the preceding section.

For example, suppose you want to change 4.51 to a fraction. The result will be a mixed number with a whole number part of 4. To find the fractional part, follow these steps:

1. **Draw a line (fraction bar) under the decimal and place a 1 underneath it.**

   Draw a line under 0.51 and place a 1 underneath it:

   $$\frac{0.51}{1}$$

2. **Move the decimal point one place to the right and add a 0 after the 1.**

   $$=\frac{5.1}{10}$$

3. **Repeat Step 2 until the decimal point moves all the way to the right so you can drop the decimal point entirely.**

   In this case, you have only one additional step:

   $$=\frac{51}{100}$$

So the mixed-number equivalent of 4.51 is $4\frac{51}{100}$.

Changing fractions to decimals

Converting fractions to decimals isn't difficult, but to do it, you need to know about decimal division. If you need to get up to speed on this, check out "Dividing decimals," earlier in this chapter.

To convert a fraction to a decimal, follow these steps:

1. **Set up the fraction as a decimal division, dividing the numerator (top number) by the denominator (bottom number).**
2. **Attach enough trailing zeros to the numerator so that you can continue dividing until you find that the answer is either a *terminating decimal* or a *repeating decimal.***

Don't worry, I explain terminating and repeating decimals later.

The last step: Terminating decimals

Sometimes when you divide the numerator of a fraction by the denominator, the division eventually works out evenly. The result is a *terminating decimal.*

For example, suppose you want to change the fraction $\frac{2}{5}$ to a decimal. Here's your first step:

$$5\overline{)2}$$

One glance at this problem, and it looks like you're doomed from the start because 5 doesn't go into 2. But watch what happens when I add a few trailing zeros. Notice that I also place another decimal point in the answer just above the first decimal point. This step is important — you can read more about it in "Dividing decimals":

$$\begin{array}{r} . \\ 5\overline{)2.000} \end{array}$$

Now you can divide because, although 5 doesn't go into 2, 5 does go into 20 four times:

$$\begin{array}{r} 0.4 \\ 5\overline{)2.000} \\ \underline{20} \\ 0 \end{array}$$

You're done! As it turns out, you needed only one trailing zero, so you can ignore the rest:

$$\frac{2}{5}=0.4$$

Because the division worked out evenly, the answer is an example of a *terminating decimal.*

As another example, suppose you want to find out how to represent $\frac{7}{16}$ as a decimal. As earlier, I attach three trailing zeros:

$$\begin{array}{r} 0.437 \\ 16\overline{)7.000} \\ \underline{64} \\ 60 \\ \underline{48} \\ 120 \\ \underline{112} \\ 8 \end{array}$$

This time, three trailing zeros aren't enough to get my answer, so I attach a few more and continue:

$$\begin{array}{r} 0.4375 \\ 16\overline{)7.000000} \\ \underline{64} \\ 60 \\ \underline{48} \\ 120 \\ \underline{112} \\ 80 \\ \underline{80} \\ 0 \end{array}$$

At last, the division works out evenly, so again the answer is a terminating decimal. Therefore, $\frac{7}{16}=0.4375$.

The endless ride: Repeating decimals

Sometimes when you try to convert a fraction to a decimal, the division *never* works out evenly. The result is a *repeating decimal* — a decimal that cycles through the same number pattern forever.

You may recognize these pesky little critters from your calculator, when an apparently simple division problem produces a long string of numbers.

For example, to change $\frac{2}{3}$ to a decimal, begin by dividing 2 by 3. As in the last section, start by adding three trailing zeros, and see where it leads:

$$\begin{array}{r} 0.666 \\ 3\overline{)2.000} \\ \underline{18} \\ 20 \\ \underline{18} \\ 20 \\ \underline{18} \\ 2 \end{array}$$

At this point, you still haven't found an exact answer. But you may notice that a repeating pattern has developed in the division. No matter how many trailing zeros you attach to the number 2, the same pattern continues forever. This answer, 0.666 … , is an example of a repeating decimal. You can write $\frac{2}{3}$ as

$$\frac{2}{3} = 0.\bar{6}$$

The bar over the 6 means that, in this decimal, the number 6 repeats forever. You can represent many simple fractions as repeating decimals. In fact, *every* fraction can be represented either as a repeating decimal or as a terminating decimal — that is, as an ordinary decimal that ends.

Now suppose you want to find the decimal representation of $\frac{5}{11}$. Here's how this problem plays out:

$$\begin{array}{r} 0.4545 \\ 11\overline{)5.0000} \\ \underline{44} \\ 60 \\ \underline{55} \\ 50 \\ \underline{44} \\ 60 \\ \underline{55} \\ 5 \end{array}$$

This time, the pattern repeats every other number — 4, then 5, then 4 again, and then 5 again, forever. Attaching more trailing zeros to the original decimal only strings out this pattern indefinitely. So you can write

$$\frac{5}{11} = 0.\overline{45}$$

This time, the bar is over both the 4 and the 5, telling you that these two numbers alternate forever.

Repeating decimals are an oddity, but they aren't hard to work with. In fact, as soon as you can show that a decimal division is repeating, you've found your answer. Just remember to place the bar only over the numbers that keep on repeating.

Some decimals never end and never repeat. You can't write them as fractions, so mathematicians have agreed on some shorter ways of naming them so that writing them out doesn't take, well, forever.

Chapter 12

Playing with Percents

In This Chapter

- Understanding what percents are
- Converting percents back and forth between decimals and fractions
- Solving both simple and difficult percent problems
- Using equations to solve three different types of percent problems

Like whole numbers and decimals, percents are a way to talk about parts of a whole. The word *percent* means "out of 100." So if you have 50% of something, you have 50 out of 100. If you have 25% of it, you have 25 out of 100. Of course, if you have 100% of anything, you have all of it.

In this chapter, I show you how to work with percents. Because percents resemble decimals, I first show you how to convert numbers back and forth between percents and decimals. No worries — this switch is easy to do. Next, I show you how to convert back and forth between percents and fractions — also not too bad. When you understand how conversions work, I show you the three basic types of percent problems, plus a method that makes the problems simple.

Making Sense of Percents

The word *percent* literally means "for 100," but in practice, it means closer to "out of 100." For example, suppose that a school has exactly 100 children — 50 girls and 50 boys. You can say that "50 out of 100" children are girls — or you can shorten it to simply "50 percent." Even shorter than that, you can use the symbol %, which means *percent.*

Saying that 50% of the students are girls is the same as saying that $\frac{1}{2}$ of them are girls. Or if you prefer decimals, it's the same thing as saying that 0.5 of all the students are girls. This example shows you that percents, like fractions and decimals, are just another way of talking about parts of the whole. In this case, the whole is the total number of children in the school.

You don't literally have to have 100 of something to use a percent. You probably won't ever really cut a cake into 100 pieces, but that doesn't matter. The values are the same. Whether you're talking about cake, a dollar, or a group of children, 50% is still half, 25% is still one-quarter, 75% is still three-quarters, and so on.

Any percentage smaller than 100% means less than the whole — the smaller the percentage, the less you have. You probably know this fact well from the school grading system. If you get 100%, you get a perfect score. And 90% is usually A work, 80% is a B, 70% is a C, and, well, you know the rest.

Of course, 0% means "0 out of 100" — any way you slice it, you have nothing.

Dealing with Percents Greater than 100%

100% means "100 out of 100" — in other words, everything. So when I say I have 100% confidence in you, I mean that I have complete confidence in you.

What about percentages more than 100%? Well, sometimes percentages like these don't make sense. For example, you can't spend more than 100% of your time playing basketball, no matter how much you love the sport; 100% is all the time you have, and there ain't no more.

But a lot of times, percentages larger than 100% are perfectly reasonable. For example, suppose I own a hot dog wagon and sell the following:

10 hot dogs in the morning

30 hot dogs in the afternoon

The number of hot dogs I sell in the afternoon is 300% of the number I sold in the morning. It's three times as many.

Here's another way of looking at this: I sell 20 more hot dogs in the afternoon than in the morning, so this is a *200% increase* in the afternoon — 20 is twice as many as 10.

Spend a little time thinking about this example until it makes sense. You visit some of these ideas again in Chapter 13, when I show you how to do word problems involving percents.

Converting to and from Percents, Decimals, and Fractions

To solve many percent problems, you need to change the percent to either a decimal or a fraction. Then you can apply what you know about solving decimal and fraction problems. For this reason, I show you how to convert to and from percents before I show you how to solve percent problems.

Percents and decimals are similar ways of expressing parts of a whole. This similarity makes converting percents to decimals, and vice versa, mostly a matter of moving the decimal point. It's so simple you can probably do it in your sleep (but you should probably stay awake when you first read about the concept).

Percents and fractions both express the same idea — parts of a whole — in different ways. So converting back and forth between percents and fractions isn't quite as simple as just moving the decimal point back and forth. In this section, I cover the ways to convert to and from percents, decimals, and fractions, starting with percents to decimals.

Going from percents to decimals

To convert a percent to a decimal, drop the percent sign (%) and move the decimal point two places to the left. It's simple. Remember that, in a whole number, the decimal point comes at the end. For example,

$$2.5\% = 0.025$$
$$4\% = 0.04$$
$$36\% = 0.36$$
$$111\% = 1.11$$

Changing decimals into percents

REMEMBER

To convert a decimal to a percent, move the decimal point two places to the right and add a percent sign (%):

$$0.07 = 7\%$$
$$0.21 = 21\%$$
$$0.375 = 37.5\%$$

Switching from percents to fractions

Converting percents to fractions is fairly straightforward. Remember that the word percent means "out of 100." So changing percents to fractions naturally involves the number 100.

To convert a percent to a fraction, use the number in the percent as your numerator (top number) and the number 100 as your denominator (bottom number):

$$39\% = \frac{39}{100} \quad 86\% = \frac{86}{100} \quad 217\% = \frac{217}{100}$$

As always with fractions, you may need to reduce to lowest terms or convert an improper fraction to a mixed number (flip to Chapter 9 for more on these topics).

In the three examples, $\frac{39}{100}$ can't be reduced or converted to a mixed number. However, $\frac{86}{100}$ can be reduced because the numerator and denominator are both even numbers:

$$\frac{86}{100} = \frac{43}{50}$$

And $\frac{217}{100}$ can be converted to a mixed number because the numerator (217) is greater than the denominator (100):

$$\frac{217}{100} = 2\frac{17}{100}$$

Once in a while, you may start out with a percentage that's a decimal, such as 99.9%. The rule is still the same, but now you have a decimal in the numerator (top number), which most people don't like to see. To get rid of it, move the decimal point one place to the right in both the numerator and the denominator:

$$99.9\% = \frac{99.9}{100} = \frac{999}{1000}$$

Thus, 99.9% converts to the fraction $\frac{999}{1000}$.

Turning fractions into percents

Converting a fraction to a percent is really a two-step process. Here's how to convert a fraction to a percent:

1. **Convert the fraction to a decimal.**

   For example, suppose you want to convert the fraction $\frac{4}{5}$ to a percent. To convert $\frac{4}{5}$ to a decimal, you can divide the numerator by the denominator, as shown in Chapter 11:

   $$\frac{4}{5} = 0.8$$

2. **Convert this decimal to a percent.**

   Convert 0.8 to a percent by moving the decimal point two places to the right and adding a percent sign (as I show you earlier in "Changing decimals into percents").

   $$0.8 = 80\%$$

Now suppose you want to convert the fraction $\frac{5}{8}$ to a percent. Follow these steps:

1. **Convert $\frac{5}{8}$ to a decimal by dividing the numerator by the denominator:**

   $$\begin{array}{r} 0.625 \\ 8\overline{)5.000} \\ \underline{48} \\ 20 \\ \underline{16} \\ 40 \\ \underline{40} \\ 0 \end{array}$$

   Therefore, $\frac{5}{8} = 0.625$.

2. **Convert 0.625 to a percent by moving the decimal point two places to the right and adding a percent sign (%):**

   $$0.625 = 62.5\%$$

Solving Percent Problems

When you know the connection between percents and fractions, which I discuss earlier in "Converting to and from Percents, Decimals, and Fractions," you can solve a lot of percent problems with a few simple tricks. Other problems, however, require a bit more work. In this section, I show you how to tell an easy percent problem from a tough one, and I give you the tools to solve all of them.

Figuring out simple percent problems

A lot of percent problems turn out to be easy when you give them a little thought. In many cases, just remember the connection between percents and fractions, and you're halfway home:

- **Finding 100% of a number:** Remember that 100% means the whole thing, so 100% of any number is simply the number itself:

  100% of 5 is 5.

  100% of 91 is 91.

  100% of 732 is 732.

- **Finding 50% of a number:** Remember that 50% means half, so to find 50% of a number, just divide it by 2:

  50% of 20 is 10.

  50% of 88 is 44.

  50% of 7 is $\frac{7}{2}$ (or $3\frac{1}{2}$ or 3.5).

- **Finding 25% of a number:** Remember that 25% equals $\frac{1}{4}$, so to find 25% of a number, divide it by 4:

  $$25\% \text{ of } 40 = 10$$
  $$25\% \text{ of } 88 = 22$$
  $$25\% \text{ of } 15 = \frac{15}{4} = 3\frac{3}{4} = 3.75$$

- **Finding 20% of a number:** Finding 20% of a number is handy if you like the service you've received in a restaurant, because a good tip is 20% of the check. Because 20% equals $\frac{1}{5}$, you can find 20% of a number by dividing it by 5. But I can show you an easier way: Remember that 20% is

2 times 10%, so to find 20% of a number, move the decimal point onc place to the left and double the result:

$$20\% \text{ of } 80 = 8 \times 2 = 16$$
$$20\% \text{ of } 300 = 30 \times 2 = 60$$
$$20\% \text{ of } 41 = 4.1 \times 2 = 8.2$$

- **Finding 10% of a number:** Finding 10% of any number is the same as finding $\frac{1}{10}$ of that number. To do this, just move the decimal point one place to the left:

$$10\% \text{ of } 30 = 3$$
$$10\% \text{ of } 41 = 4.1$$
$$10\% \text{ of } 7 = 0.7$$

- **Finding 200%, 300%, and so on of a number:** Working with percents that are multiples of 100 is easy. Just drop the two 0s and multiply by the number that's left:

$$200\% \text{ of } 7 = 2 \times 7 = 14$$
$$300\% \text{ of } 10 = 3 \times 10 = 30$$
$$1{,}000\% \text{ of } 45 = 10 \times 45 = 450$$

(See the earlier "Dealing with Percents Greater than 100%" section for details on what having more than 100% really means.)

Turning the problem around

Here's a trick that makes certain tough-looking percent problems so easy that you can do them in your head. Simply move the percent sign from one number to the other and flip the order of the numbers.

Suppose someone wants you to figure out the following:

$$88\% \text{ of } 50$$

Finding 88% of anything isn't an activity anybody looks forward to. But an easy way of solving the problem is to switch it around:

$$88\% \text{ of } 50 = 50\% \text{ of } 88$$

This move is perfectly valid, and it makes the problem a lot easier. It works because the word *of* really means multiplication, and you can multiply either backward or forward and get the same answer. As I discuss in the preceding section, "Figuring out simple percent problems," 50% of 88 is simply half of 88:

$$88\% \text{ of } 50 = 50\% \text{ of } 88 = 44$$

As another example, suppose you want to find

$$7\% \text{ of } 200$$

Again, finding 7% is tricky, but finding 200% is simple, so switch the problem around:

$$7\% \text{ of } 200 = 200\% \text{ of } 7$$

In the preceding section, I tell you that, to find 200% of any number, you just multiply that number by 2:

$$7\% \text{ of } 200 = 200\% \text{ of } 7 = 2 \times 7 = 14$$

Deciphering more-difficult percent problems

You can solve a lot of percent problems, using the tricks I show you earlier in this chapter. For more difficult problems, you may want to switch to a calculator. If you don't have a calculator at hand, solve percent problems by turning them into decimal multiplication, as follows:

1. **Change the word *of* to a multiplication sign and the percent to a decimal (as I show you earlier in this chapter).**

   Suppose you want to find 35% of 80. Here's how you start:

   $$35\% \text{ of } 80 = 0.35 \times 80$$

2. **Solve the problem using decimal multiplication (see Chapter 11).**

   Here's what the example looks like:

   $$\begin{array}{r} 0.35 \\ \times\ 80 \\ \hline 28.00 \end{array}$$

   So 35% of 80 is 28.

Putting All the Percent Problems Together

In the preceding section, "Solving Percent Problems," I give you a few ways to find any percent of any number. This type of percent problem is the most common, which is why it gets top billing.

But percents crop up in a wide range of business applications, such as banking, real estate, payroll, and taxes. (I show you some real-world applications when I discuss word problems in Chapter 13.) And depending on the situation, two other common types of percent problems may present themselves.

In this section, I show you these two additional types of percent problems and how they relate to the type you now know how to solve. I also give you a simple tool to make quick work of all three types.

Identifying the three types of percent problems

Earlier in this chapter, I show you how to solve problems that look like this:

50% of 2 is ?

The answer, of course, is 1. (See "Solving Percent Problems" for details on how to get this answer.) Given two pieces of information — the percent and the number to start with — you can figure out what number you end up with.

Now suppose instead that I leave out the percent but give you the starting and ending numbers:

? % of 2 is 1

You can still fill in the blank without too much trouble. Similarly, suppose that I leave out the starting number but give the percent and the ending number:

50% of ? is 1

Again, you can fill in the blank.

If you get this basic idea, you're ready to solve percent problems. When you boil them down, nearly all percent problems are like one of the three types I show in Table 12-1.

Table 12-1 The Three Main Types of Percent Problems

Problem Type	*What to Find*	*Example*
Type #1	The ending number	50% of 2 is *what*?
Type #2	The percentage	*What* percent of 2 is 1?
Type #3	The starting number	50% of *what* is 1?

In each case, the problem gives you two of the three pieces of information, and your job is to figure out the remaining piece. In the next section, I give you a simple tool to help you solve all three of these types of percent problems.

Solving percent problems with equations

Here's how to solve any percent problem:

1. **Change the word *of* to a multiplication sign and the percent to a decimal (as I show you earlier in this chapter).**

   This step is the same as for more straightforward percent problems. For example, consider this problem:

   60% of what is 75?

   Begin by changing as follows:

60%	of	what	is	75
0.6	$\times$			75

2. **Turn the word *is* to an equals sign and the word *what* into the letter *n*.**

   Here's what this step looks like:

60%	of	what	is	75
0.6	$\times$	n	$=$	75

   This equation looks more normal, as follows:

   $$0.6 \times n = 75$$

3. **Find the value of *n*.**

   Technically, the last step involves a little bit of algebra, but I know you can handle it. (For a complete explanation of algebra, see Part V of this book.) In the equation, n is being multiplied by 0.6. You want to "undo" this operation by *dividing* by 0.6 on both sides of the equation:

$$0.6 \times n \div 0.6 = 75 \div 0.6$$

Almost magically, the left side of the equation becomes a lot easier to work with because multiplication and division by the same number cancel each other out:

$$n = 75 \div 0.6$$

Remember that n is the answer to the problem. If your teacher lets you use a calculator, this last step is easy; if not, you can calculate it using some decimal division, as I show you in Chapter 11:

$$n = 125$$

Either way, the answer is 125 — so 60% of 125 is 75.

As another example, suppose you're faced with this percent problem:

What percent of 250 is 375?

To begin, change the *of* into a multiplication sign and the percent into a decimal.

What	percent	of	250	is	375
	× 0.01	×	250		375

Notice here that, because I don't know the percent, I change the word *percent* to × 0.01. Next, change *is* to an equals sign and *what* to the letter n:

What	percent	of	250	is	375
n	× 0.01	×	250	=	375

Consolidate the equation and then multiply:

$$n \times 2.5 = 375$$

Now divide both sides by 2.5:

$$n = 375 \div 2.5 = 150$$

Therefore, the answer is 150 — so 150% of 250 is 375.

Here's one more problem: 49 is what percent of 140? Begin, as always, by translating the problem into words:

49	is	what	percent	of	140
49	$=$	n	$\times 0.01$	$\times$	140

Simplify the equation:

$$49 = n \times 1.4$$

Now divide both sides by 1.4:

$$49 \div 1.4 = n \times 1.4 \div 1.4$$

Again, multiplication and division by the same number allows you to cancel on the left side of the equation and complete the problem:

$$49 \div 1.4 = n$$

$$35 = n$$

Therefore, the answer is 35, so 49 is 35% of 140.

Chapter 13

Word Problems with Fractions, Decimals, and Percents

In This Chapter

- Adding and subtracting fractions, decimals, and percents in word equations
- Translating the word *of* as multiplication
- Changing percents to decimals in word problems
- Tackling business problems involving percent increase and decrease

In Chapter 6, I show you how to solve word problems (also known as story problems) by setting up word equations that use the Big Four operations (adding, subtracting, multiplying, and dividing). In this chapter, I show you how to extend these skills to solve word problems with fractions, decimals, and percents.

First, I show you how to solve relatively easy problems, in which all you need to do is add or subtract fractions, decimals, or percents. Next, I show you how to solve problems that require you to multiply fractions. Such problems are easy to spot because they almost always contain the word *of*. After that, you discover how to solve percent problems by setting up a word equation and changing the percent to a decimal. Finally, I show you how to handle problems of percent increase and decrease. These problems are often practical money problems in which you figure out information about raises and salaries, costs and discounts, or amounts before and after taxes.

Adding and Subtracting Parts of the Whole in Word Problems

Certain word problems involving fractions, decimals, and percents are really just problems in adding and subtracting. You may add fractions, decimals, or percents in a variety of real-world settings that rely on weights and measures — such as cooking and carpentry. (In Chapter 15, I discuss these applications in depth.)

To solve these problems, you can use the skills that you pick up in Chapters 10 (for adding and subtracting fractions), 11 (for adding and subtracting decimals), and 12 (for adding and subtracting percents).

Sharing a pizza: Fractions

You may have to add or subtract fractions in problems that involve splitting up part of a whole. For example, consider the following:

> Joan ate $\frac{1}{6}$ of a pizza, Tony ate $\frac{1}{4}$, and Sylvia ate $\frac{1}{3}$. What fraction of the pizza was left when they were finished?

In this problem, just jot down the information that's given as word equations:

$$\text{Joan} = \frac{1}{6} \quad \text{Tony} = \frac{1}{4} \quad \text{Sylvia} = \frac{1}{3}$$

These fractions are part of one total pizza. To solve the problem, you need to find out how much all three people ate, so form the following word equation:

all three = Joan + Tony + Sylvia

Now you can substitute as follows:

$$\text{all three} = \frac{1}{6} + \frac{1}{4} + \frac{1}{3}$$

Chapter 10 gives you several ways to add these fractions. Here's one way:

$$\text{all three} = \frac{2}{12} + \frac{3}{12} + \frac{4}{12} = \frac{9}{12} = \frac{3}{4}$$

However, the question asks what fraction of the pizza was left after they finished, so you have to subtract that amount from the whole:

$$1 - \frac{3}{4} = \frac{1}{4}$$

Thus, the three people left $\frac{1}{4}$ of a pizza.

Buying by the pound: Decimals

You frequently work with decimals when dealing with money, metric measurements (see Chapter 15), and food sold by the pound. The following problem requires you to add and subtract decimals, which I discuss in Chapter 11. Even though the decimals may look intimidating, this problem is fairly simple to set up:

> Antonia bought 4.53 pounds of beef and 3.1 pounds of lamb. Lance bought 5.24 pounds of chicken and 0.7 pounds of pork. Which of them bought more meat, and how much more?

To solve this problem, you first find out how much each person bought:

$$\text{Antonia} = 4.53 + 3.1 = 7.63$$
$$\text{Lance} = 5.24 + 0.7 = 5.94$$

You can already see that Antonia bought more than Lance. To find how much more, subtract:

$$7.63 - 5.94 = 1.69$$

So Antonia bought 1.69 pounds more than Lance.

Splitting the vote: Percents

When percents represent answers in polls, votes in an election, or portions of a budget, the total often has to add up to 100%. In real life, you may see such info organized as a pie chart (which I discuss in Chapter 17). Solving problems about this kind of information often involves nothing more than adding and subtracting percents. Here's an example:

> In a recent mayoral election, five candidates were on the ballot. Faber won 39% of the vote, Gustafson won 31%, Ivanovich won 18%, Dixon won 7%, Obermayer won 3%, and the remaining votes went to write-in candidates. What percentage of voters wrote in their selection?

The candidates were in a single election, so all the votes have to total 100%. The first step here is just to add up the five percentages. Then subtract that value from 100%:

$$39\% + 31\% + 18\% + 7\% + 3\% = 98\%$$
$$100\% - 98\% = 2\%$$

Because 98% of voters voted for one of the five candidates, the remaining 2% wrote in their selections.

Problems about Multiplying Fractions

In word problems, the word *of* almost always means multiplication. So whenever you see the word *of* following a fraction, decimal, or percent, you can usually replace it with a times sign.

When you think about it, *of* means multiplication even when you're not talking about fractions. For example, when you point to an item in a store and say, "I'll take three of those," in a sense you're saying, "I'll take that one multiplied by three."

The following examples give you practice turning word problems that include the word *of* into multiplication problems that you can solve with fraction multiplication.

Renegade grocery shopping: Buying less than they tell you to

When you understand that the word *of* means multiplication, you have a powerful tool for solving word problems. For instance, you can figure out how much you'll spend if you don't buy food in the quantities listed on the signs. Here's an example:

If beef costs $4 a pound, how much does $\frac{5}{8}$ of a pound cost?

Here's what you get if you simply change the *of* to a multiplication sign:

$$\frac{5}{8} \times 1 \text{ pound of beef}$$

So you know how much beef you're buying. However, you want to know the cost. Because the problem tells you that 1 pound = $4, you can replace 1 pound of beef with $4:

$$= \frac{5}{8} \times \$4$$

Now you have an expression you can evaluate. Use the rules of multiplying fractions from Chapter 10 and solve:

$$= \frac{5 \times \$4}{8} = \$\frac{20}{8}$$

This fraction reduces to $\$\frac{5}{2}$. However, the answer looks weird because dollars are usually expressed in decimals, not fractions. So convert this fraction to a decimal using the rules I show you in Chapter 11:

$$\$\frac{5}{2} = \$2.5 = \$2.50$$

At this point, recognize that \$2.5 is more commonly written as \$2.50, and you have your answer.

Easy as pie: Working out what's left on your plate

Sometimes when you're sharing something such as a pie, not everyone gets to it at the same time. The eager pie-lovers snatch the first slice, not bothering to divide the pie into equal servings, and the people who were slower, more patient, or just not that hungry cut their own portions from what's left over. When someone takes a part of the leftovers, you can do a bit of multiplication to see how much of the whole pie that portion represents.

Consider the following example:

> Jerry bought a pie and ate $\frac{1}{5}$ of it. Then his wife, Doreen, ate $\frac{1}{6}$ of what was left. How much of the total pie was left?

To solve this problem, begin by jotting down what the first sentence tells you:

$$\text{Jerry} = \frac{1}{5}$$

Doreen ate part of what was left, so write a word equation that tells you how much of the pie was left after Jerry was finished. He started with a whole pie, so subtract his portion from 1:

$$\text{pie left after Jerry} = 1 - \frac{1}{5} = \frac{4}{5}$$

Next, Doreen ate $\frac{1}{6}$ of this amount. Rewrite the word *of* as multiplication and solve as follows. This answer tells you how much of the whole pie Doreen ate:

$$\text{Doreen} = \frac{1}{6} \times \frac{4}{5} = \frac{4}{30}$$

To make the numbers a little smaller before you go on, notice that you can reduce the fraction:

$$\text{Doreen} = \frac{2}{15}$$

Now you know how much Jerry and Doreen both ate, so you can add these amounts together:

$$\text{Jerry} + \text{Doreen} = \frac{1}{5} + \frac{2}{15}$$

Solve this problem as I show you in Chapter 10:

$$= \frac{3}{15} + \frac{2}{15} = \frac{5}{15}$$

This fraction reduces to $\frac{1}{3}$. Now you know that Jerry and Doreen ate $\frac{1}{3}$ of the pie, but the problem asks you how much is left. So finish up with some subtraction and write the answer:

$$1 - \frac{1}{3} = \frac{2}{3}$$

The amount of pie left over was $\frac{2}{3}$.

Multiplying Decimals and Percents in Word Problems

In the preceding section, "Problems about Multiplying Fractions," I show you how the word *of* in a fraction word problem usually means multiplication. This idea is also true in word problems involving decimals and percents.

The method for solving these two types of problems is similar, so I lump them together in this section.

You can easily solve word problems involving percents by changing the percents into decimals (see Chapter 12 for details). Here are a few common percents and their decimal equivalents:

$25\% = 0.25$ $50\% = 0.5$ $75\% = 0.75$ $99\% = 0.99$

To the end: Figuring out how much money is left

One common type of problem gives you a starting amount — and a bunch of other information — and then asks you to figure out how much you end up with. Here's an example:

> Maria's grandparents gave her $125 for her birthday. She put 40% of the money in the bank, spent 35% of what was left on a pair of shoes, and then spent the rest on a dress. How much did the dress cost?

Start at the beginning, forming a word equation to find out how much money Maria put in the bank:

money in bank = 40% of $125

To solve this word equation, change the percent to a decimal and the word *of* to a multiplication sign; then multiply:

money in bank = 0.4 × $125 = $50

Pay special attention to whether you're calculating how much of something was used up or how much of something is left over. If you need to work with the portion that remains, you may have to subtract the amount used from the amount you started with.

Because Maria started with $125, she had $75 left to spend:

money left to spend

= money from grandparents – money in bank

= $125 – $50

= $75

The problem then says that she spent 35% of this amount on a pair of shoes. Again, change the percent to a decimal and the word *of* to a multiplication sign:

$$\text{shoes} = 35\% \text{ of } \$75 = 0.35 \times \$75 = \$26.25$$

She spent the rest of the money on a dress, so

$$\text{dress} = \$75 - \$26.25 = \$48.75$$

Therefore, Maria spent $48.75 on the dress.

Finding out how much you started with

Some problems give you the amount that you end up with and ask you to find out how much you started with. In general, these problems are harder because you're not used to thinking backward. Here's an example, and it's kind of a tough one, so fasten your seat belt:

> Maria received some birthday money from her aunt. She put her usual 40% in the bank and spent 75% of the rest on a purse. When she was done, she had $12 left to spend on dinner. How much did her aunt give her?

This problem is similar to the one in the preceding section, but you need to start at the end and work backward. Notice that the only dollar amount in the problem comes after the two percent amounts. The problem tells you that she ends up with $12 after two transactions — putting money in the bank and buying a purse — and asks you to find out how much she started with.

To solve this problem, set up two word equations to describe the two transactions:

$$\text{money from aunt} - \text{money for bank} = \text{money after bank}$$
$$\text{money after bank} - \text{money for purse} = \$12$$

Notice what these two word equations are saying. The first tells you that Maria took the money from her aunt, subtracted some money to put in the bank, and left the bank with a new amount of money, which I'm calling *money after bank.* The second word equation starts where the first leaves off. It tells you that Maria took the money left over from the bank, subtracted some money for a purse, and ended up with $12.

This second equation already has an amount of money filled in, so start here. To solve this problem, realize that Maria spent 75% of her money *at that time* on the purse — that is, 75% of the money she still had after the bank:

$$\text{money after bank} - 75\% \text{ of money after bank} = \$12$$

I'm going to make one small change to this equation so you can see what it's really saying:

$$100\% \text{ of money after bank} - 75\% \text{ of money after bank} = \$12$$

Adding *100% of* doesn't change the equation because it really just means you're multiplying by 1. In fact, you can slip these two words in anywhere without changing what you mean, though you may sound ridiculous saying "Last night, I drove 100% of my car home from work, walked 100% of my dog, then took 100% of my wife to see 100% of a movie."

In this particular case, however, these words help you to make a connection because 100% – 75% = 25%; here's an even better way to write this equation:

$$25\% \text{ of money after bank} = \$12$$

Before moving on, make sure you understand the steps that have brought you here.

You know now that 25% of money after bank is $12, so the total amount of money after bank is 4 times this amount — that is, $48. Therefore, you can plug this number into the first equation:

$$\text{money from aunt} - \text{money for bank} = \$48$$

Now you can use the same type of thinking to solve this equation (and it goes a lot more quickly this time!). First, Maria placed 40% of the money from her aunt in the bank:

$$\text{money from aunt} - 40\% \text{ of money from aunt} = \$48$$

Again, rewrite this equation to make what it's saying clearer:

$$100\% \text{ of money from aunt} - 40\% \text{ of money from aunt} = \$48$$

Now, because 100% – 40% = 60%, rewrite it again:

$$60\% \text{ of money from aunt} = \$48$$

Thus, 0.6 × money from aunt = $48. Divide both sides of this equation by 0.6:

$$\text{money from aunt} = \$48 \div 0.6 = \$80$$

So Maria's aunt gave her $80 for her birthday.

Handling Percent Increases and Decreases in Word Problems

Word problems that involve increasing or decreasing by a percentage add a final spin to percent problems. Typical percent-increase problems involve calculating the amount of a salary plus a raise, the cost of merchandise plus tax, or an amount of money plus interest or dividend. Typical percent decrease problems involve the amount of a salary minus taxes or the cost of merchandise minus a discount.

To tell you the truth, you may have already solved problems of this kind earlier in "Multiplying Decimals and Percents in Word Problems." But people often get thrown by the language of these problems — which, by the way, is the language of business — so I want to give you some practice in solving them.

Raking in the dough: Finding salary increases

A little street smarts should tell you that the words *salary increase* or *raise* mean more money, so get ready to do some addition. Here's an example:

> Alison's salary was $40,000 last year, and at the end of the year, she received a 5% raise. What will she earn this year?

To solve this problem, first realize that Alison got a raise. So whatever she makes this year, it will be more than she made last year. The key to setting up this type of problem is to think of percent increase as "100% of last year's salary plus 5% of last year's salary." Here's the word equation:

this year's salary = 100% of last year's salary + 5% of last year's salary

Now you can just add the percentages (see the nearby sidebar for why this works):

this year's salary = 105% of last year's salary

Change the percent to a decimal and the word *of* to a multiplication sign; then fill in the amount of last year's salary:

this year's salary = 1.05 × $40,000

Now you're ready to multiply:

$$\text{this year's salary} = \$42{,}000$$

So Alison's new salary is $42,000.

Earning interest on top of interest

The word *interest* means more money. When you receive interest from the bank, you get more money. And when you pay interest on a loan, you pay more money. Sometimes people earn interest on the interest they earned earlier, which makes the dollar amounts grow even faster. Here's an example:

> Bethany placed $9,500 in a one-year CD that paid 4% interest. The next year, she rolled this over into a bond that paid 6% per year. How much did Bethany earn on her investment in those two years?

This problem involves interest, so it's another problem in percent increase — only this time, you have to deal with two transactions. Take them one at a time.

The first transaction is a percent increase of 4% on $9,500. The following word equation makes sense:

$$\begin{aligned}\text{money after first year} &= 100\% \text{ of initial deposit} + 4\% \text{ of initial deposit}\\ &= 104\% \text{ of initial deposit}\end{aligned}$$

Now, substitute $9,500 for the initial deposit and calculate:

$$\begin{aligned}&= 104\% \text{ of } \$9{,}500\\ &= 1.04 \times \$9{,}500\\ &= \$9{,}880\end{aligned}$$

At this point, you're ready for the second transaction. This is a percent increase of 6% on $9,880:

$$\begin{aligned}\text{final amount} &= 106\% \text{ of } \$9{,}880\\ &= 1.06 \times \$9{,}880\\ &= \$10{,}472.80\end{aligned}$$

Then subtract the initial deposit from the final amount:

$$\begin{aligned}\text{earnings} &= \text{final amount} - \text{initial deposit}\\ &= \$10,472.80 - \$9,500\\ &= \$972.80\end{aligned}$$

So Bethany earned $972.80 on her investment.

Getting a deal: Calculating discounts

When you hear the words *discount* or *sale price,* think of subtraction. Here's an example:

> Greg has his eye on a television with a listed price of $2,100. The salesman offers him a 30% discount if he buys it today. What will the television cost with the discount?

In this problem, you need to realize that the discount lowers the price of the television, so you have to subtract:

$$\begin{aligned}\text{Sale price} &= 100\% \text{ of regular price} - 30\% \text{ of regular price}\\ &= 70\% \text{ of regular price}\\ &= 0.7 \times \$2,100 = \$1,470\end{aligned}$$

Thus, the television costs $1,470 with the discount.

Part IV

Picturing and Measuring — Graphs, Measures, Stats, and Sets

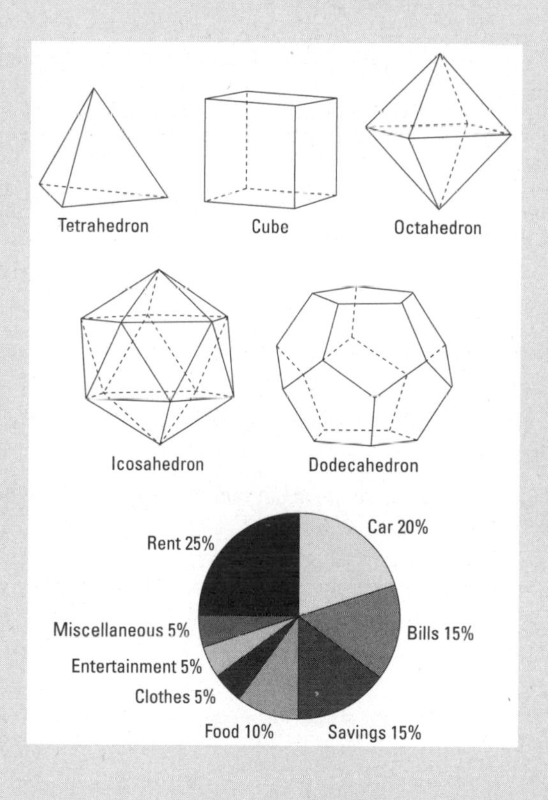

For find out how to use probability to calculate the odds in dice games, go to www.dummies.com/extras/basicmathandprealgebra

In this part...

- Represent very large and very small numbers with scientific notation
- Weigh and measure with both the English and metric systems
- Understand basic geometry, including points, lines, and angles, plus basic shapes and solids
- Present math info visually, using bar graphs, pie charts, line graphs, and the *xy*-graph
- Solve word problems involving measurement and geometry
- Answer real-world questions with statistics and probability
- Get familiar with some basic set theory, including union and intersection

Chapter 14

A Perfect Ten: Condensing Numbers with Scientific Notation

In This Chapter

- Knowing how to express powers of ten in exponential form
- Appreciating how and why scientific notation works
- Understanding order of magnitude
- Multiplying numbers in scientific notation

Scientists often work with very small or very large measurements — the distance to the next galaxy, the size of an atom, the mass of the Earth, or the number of bacteria cells growing in last week's leftover Chinese take-out. To save on time and space — and to make calculations easier — people developed a sort of shorthand called *scientific notation*.

Scientific notation uses a sequence of numbers known as the powers of ten, which I introduce in Chapter 2:

1 10 100 1,000 10,000 100,000 1,000,000 10,000,000 ...

Each number in the sequence is 10 times more than the preceding number.

Powers of ten are easy to work with, especially when you're multiplying and dividing, because you can just add or drop zeros or move the decimal point. They're also easy to represent in exponential form (as I show you in Chapter 4):

10^0 10^1 10^2 10^3 10^4 10^5 10^6 10^7 ...

Scientific notation is a handy system for writing very large and very small numbers without writing a bunch of 0s. It uses both decimals and exponents (so if you need a little brushing up on decimals, flip to Chapter 11). In this chapter, I introduce you to this powerful method of writing numbers. I also explain the order of magnitude of a number. Finally, I show you how to multiply numbers written in scientific notation.

First Things First: Using Powers of Ten as Exponents

Scientific notation uses powers of ten expressed as exponents, so you need a little background before you can jump in. In this section, I round out your knowledge of exponents, which I first introduce in Chapter 4.

Counting zeros and writing exponents

Numbers starting with a 1 and followed by only 0s (such 10, 100, 1,000, 10,000, and so forth) are called *powers of ten,* and they're easy to represent as exponents. Powers of ten are the result of multiplying 10 times itself any number of times.

To represent a number that's a power of 10 as an exponential number, count the zeros and raise 10 to that exponent. For example, 1,000 has three zeros, so $1{,}000 = 10^3$ (10^3 means to take 10 times itself three times, so it equals $10 \times 10 \times 10$). Table 14-1 shows a list of some powers of ten.

Table 14-1 Powers of Ten Expressed as Exponents

Number	*Exponent*
1	10^0
10	10^1
100	10^2
1,000	10^3
10,000	10^4
100,000	10^5
1,000,000	10^6

When you know this trick, representing a lot of large numbers as powers of ten is easy — just count the 0s! For example, the number 1 trillion — 1,000,000,000,000 — is a 1 with twelve 0s after it, so

$$1,000,000,000,000 = 10^{12}$$

This trick may not seem like a big deal, but the higher the numbers get, the more space you save by using exponents. For example, a really big number is a googol, which is 1 followed by a hundred 0s. You can write this:

10,000,000,000,000,000,000,000,000,000,000,000,000,000,000,000,000,000,000,00
0,000,000,000,000,000,000,000,000,000,000,000,000,000,000,000

As you can see, a number of this size is practically unmanageable. You can save yourself some trouble and write 10^{100}.

A 10 raised to a negative number is also a power of ten.

You can also represent decimals using negative exponents. For example,

$$10^{-1} = 0.1 \quad 10^{-2} = 0.01 \quad 10^{-3} = 0.001 \quad 10^{-4} = 0.0001$$

Although the idea of negative exponents may seem strange, it makes sense when you think about it alongside what you know about positive exponents. For example, to find the value of 10^7, start with 1 and make it larger by moving the decimal point seven spaces to the right:

$$10^7 = 10,000,000$$

Similarly, to find the value of 10^{-7}, start with 1 and make it smaller by moving the decimal point seven spaces to the left:

$$10^{-7} = 0.0000001$$

Negative powers of 10 always have one fewer 0 between the 1 and the decimal point than the power indicates. In this example, notice that 10^{-7} has six 0s between them.

As with very large numbers, using exponents to represent very small decimals makes practical sense. For example,

$$10^{-23} = 0.00000000000000000000001$$

As you can see, this decimal is easy to work with in its exponential form but almost impossible to read otherwise.

Adding exponents to multiply

An advantage of using the exponential form to represent powers of ten is that this form is a cinch to multiply. To multiply two powers of ten in exponential form, add their exponents. Here are a few examples:

- $10^1 \times 10^2 = 10^{1+2} = 10^3$

  Here, I simply multiply these numbers: $10 \times 100 = 1{,}000$
- $10^{14} \times 10^{15} = 10^{14+15} = 10^{29}$

  Here's what I'm multiplying: 100,000,000,000,000 × 1,000,000,000,000,000 = 100,000,000,000,000,000,000,000,000,000

  You can verify that this multiplication is correct by counting the 0s.
- $10^{100} \times 10^0 = 10^{100+0} = 10^{100}$

  Here I'm multiplying a googol by 1 (any number raised to an exponent of 0 equals 1), so the result is a googol.

In each of these cases, you can think of multiplying powers of ten as adding extra 0s to the number.

The rules for multiplying powers of ten by adding exponents also apply to negative exponents. For example,

$$10^3 \times 10^{-5} = 10^{3-5} = 10^{-2} = 0.01$$

Working with Scientific Notation

Scientific notation is a system for writing very large and very small numbers that makes them easier to work with. Every number can be written in scientific notation as the product of two numbers (two numbers multiplied together):

- A decimal greater than or equal to 1 and less than 10 (see Chapter 11 for more on decimals)
- A power of ten written as an exponent (see the preceding section)

Writing in scientific notation

Here's how to write any number in scientific notation:

1. **Write the number as a decimal (if it isn't one already).**

   Suppose you want to change the number 360,000,000 to scientific notation. First, write it as a decimal:

   360,000,000.0

2. **Move the decimal point just enough places to change this number to a new number that's between 1 and 10.**

   Move the decimal point to the right or left so that only one nonzero digit comes before the decimal point. Drop any leading or trailing zeros as necessary.

   Using 360,000,000.0, only the 3 should come before the decimal point. So move the decimal point eight places to the left, drop the trailing zeros, and get 3.6:

   360,000,000.0 becomes 3.6.

3. **Multiply the new number by 10 raised to the number of places you moved the decimal point in Step 2.**

   You moved the decimal point eight places, so multiply the new number by 10^8:

   3.6×10^8

4. **If you moved the decimal point to the right in Step 2, put a minus sign on the exponent.**

   You moved the decimal point to the left, so you don't have to take any action here. Thus, 360,000,000 in scientific notation is 3.6×10^8.

Changing a decimal to scientific notation basically follows the same process. For example, suppose you want to change the number 0.00006113 to scientific notation:

1. **Write 0.00006113 as a decimal (this step's easy because it's already a decimal):**

   0.00006113

2. **To change 0.00006113 to a new number between 1 and 10, move the decimal point five places to the right and drop the leading zeros:**

   6.113

3. **Because you moved the decimal point five places to the right, multiply the new number by 10^{-5}:**

$$6.113 \times 10^{-5}$$

So 0.00006113 in scientific notation is 6.113×10^{-5}.

When you get used to writing numbers in scientific notation, you can do it all in one step. Here are a few examples:

$$17{,}400 = 1.74 \times 10^4$$
$$212.04 = 2.1204 \times 10^2$$
$$0.003002 = 3.002 \times 10^{-3}$$

Seeing why scientific notation works

When you understand how scientific notation works, you're in a better position to understand why it works. Suppose you're working with the number 4,500. First of all, you can multiply any number by 1 without changing it, so here's a valid equation:

$$4{,}500 = 4{,}500 \times 1$$

Because 4,500 ends in a 0, it's divisible by 10 (see Chapter 7 for info on divisibility). So you can factor out a 10 as follows:

$$4{,}500 = 450 \times 10$$

Also, because 4,500 ends in two 0s, it's divisible by 100, so you can factor out 100:

$$4{,}500 = 45 \times 100$$

In each case, you drop another 0 after the 45 and place it after the 1. At this point, you have no more 0s to drop, but you can continue the pattern by moving the decimal point one place to the left:

$$\begin{aligned} 4{,}500 &= 4.5 \times 1{,}000 \\ &= 0.45 \times 10{,}000 \\ &= 0.045 \times 100{,}000 \end{aligned}$$

What you've been doing from the beginning is moving the decimal point one place to the left and multiplying by 10. But you can just as easily move the decimal point one place to the right and multiply by 0.1, two places right by multiplying by 0.01, and three places right by multiplying by 0.001:

$$\begin{aligned} 4{,}500 &= 45{,}000 \times 0.1 \\ &= 450{,}000 \times 0.01 \\ &= 4{,}500{,}0000 \times 0.001 \end{aligned}$$

As you can see, you have total flexibility to express 4,500 as a decimal multiplied by a power of ten. As it happens, in scientific notation, the decimal must be between 1 and 10, so the following form is the equation of choice:

$$4{,}500 = 4.5 \times 1{,}000$$

The final step is to change 1,000 to exponential form. Just count the 0s in 1,000 and write that number as the exponent on the 10:

$$4{,}500 = 4.5 \times 10^3$$

The net effect is that you moved the decimal point three places to the left and raised 10 to an exponent of 3. You can see how this idea can work for any number, no matter how large or small.

Understanding order of magnitude

A good question to ask is why scientific notation always uses a decimal between 1 and 10. The answer has to do with order of magnitude. *Order of magnitude* is a simple way to keep track of roughly how large a number is so you can compare numbers more easily. The order of magnitude of a number is its exponent in scientific notation. For example,

$$\begin{aligned} 703 &= 7.03 \times 10^2 \text{ (order of magnitude is } 2) \\ 600{,}000 &= 6 \times 10^5 \text{ (order of magnitude is } 5) \\ 0.00095 &= 9.5 \times 10^{-4} \text{ (order of magnitude is } -4) \end{aligned}$$

Every number starting with 10 but less than 100 has an order of magnitude of 1. Every number starting with 100 but less than 1,000 has an order of magnitude of 2.

Multiplying with scientific notation

Multiplying numbers that are in scientific notation is fairly simple because multiplying powers of ten is easy, as you see earlier in this chapter in "Adding exponents to multiply." Here's how to multiply two numbers that are in scientific notation:

1. **Multiply the two decimal parts of the numbers.**

   Suppose you want to multiply the following:

   $$(4.3 \times 10^5)(2 \times 10^7)$$

   Multiplication is commutative (see Chapter 4), so you can change the order of the numbers without changing the result. And because of the associative property, you can also change how you group the numbers. Therefore, you can rewrite this problem as

   $$(4.3 \times 2)(10^5 \times 10^7)$$

   Multiply what's in the first set of parentheses — 4.3×2 — to find the decimal part of the solution:

   $$4.3 \times 2 = 8.6$$

2. **Multiply the two exponential parts by adding their exponents.**

   Now multiply $10^5 \times 10^7$:

   $$10^5 \times 10^7 = 10^{5+7} = 10^{12}$$

3. **Write the answer as the product of the numbers you found in Steps 1 and 2.**

   $$8.6 \times 10^{12}$$

4. **If the decimal part of the solution is 10 or greater, move the decimal point one place to the left and add 1 to the exponent.**

   Because 8.6 is less than 10, you don't have to move the decimal point again, so the answer is 8.6×10^{12}.

   Note: This number equals 8,600,000,000,000.

This method works even when one or both of the exponents are negative numbers. For example, if you follow the preceding series of steps, you find that $(6.02 \times 10^{23})(9 \times 10^{-28}) = 5.418 \times 10^{-4}$. ***Note:*** In decimal form, this number equals 0.0005418.

Chapter 15

How Much Have You Got? Weights and Measures

In This Chapter

- Using units for non-discrete measurement
- Discovering differences between the English and metric systems
- Estimating and calculating English and metric system conversions

In Chapter 4, I introduce you to *units,* which are items that can be counted, such as apples, coins, or hats. Apples, coins, and hats are easy to count because they're *discrete* — that is, you can easily see where one ends and the next one begins. But not everything is so easy. For example, how do you count water — by the drop? Even if you tried, exactly how big is a drop?

Units of measurement come in handy at this point. A *unit of measurement* allows you to count something that isn't discrete: an amount of a liquid or solid, the distance from one place to another, a length of time, the speed at which you're traveling, or the temperature of the air.

In this chapter, I discuss two important systems of measurement: English and metric. You're probably familiar with the English system already, and you may know more than you think about the metric system. Each of these measurement systems provides a different way to measure distance, volume, weight (or mass), time, and speed. Next, I show you how to estimate metric amounts in English units. Finally, I show how to convert from English units to metric and vice versa.

Examining Differences between the English and Metric Systems

The two most common measurement systems today are the *English system* and the *metric system.*

Most Americans learn the units of the English system — for example, pounds and ounces, feet and inches, and so forth — and use them every day. Unfortunately, the English system is awkward for use with math. English units such as inches and fluid ounces are often measured in fractions, which (as you may know from Chapters 9 and 10) can be difficult to work with.

The *metric system* was invented to simplify the application of math to measurement. Metric units are based on the number 10, which makes them much easier to work with. Parts of units are expressed as decimals, which (as Chapter 11 shows you) are much friendlier than fractions.

Yet despite these advantages, the metric system has been slow to catch on in the U.S. Many Americans feel comfortable with English units and are reluctant to part with them. For example, if I ask you to carry a 20-lb. bag for one-fourth of a mile, you know what to expect. However, if I ask you to carry a bag weighing 10 kilograms half a kilometer, you may not be sure.

In this section, I show you the basic units of measurement for both the English and metric systems.

If you want an example of the importance of converting carefully, you may want to look to NASA — they kind of lost a Mars orbiter in the late 1990s because an engineering team used English units and NASA used metric to navigate!

Looking at the English system

The *English system of measurement* is most commonly used in the United States (but, ironically, not in England). Although you're probably familiar with most of the English units of measurement, in the following list, I make sure you know the most important ones. I also show you some equivalent values that can help you do conversions from one type of unit to another.

- **Units of distance:** Distance — also called *length* — is measured in inches (in.), feet (ft.), yards (yd.), and miles (mi.):

  $$12 \text{ inches} = 1 \text{ foot}$$
  $$3 \text{ feet} = 1 \text{ yard}$$
  $$5{,}280 \text{ feet} = 1 \text{ mile}$$

- **Units of fluid volume:** Fluid volume (also called *capacity*) is the amount of space occupied by a liquid, such as water, milk, or wine. I discuss volume when I talk about geometry in Chapter 16. Volume is measured in fluid ounces (fl. oz.), cups (c.), pints (pt.), quarts (qt.), and gallons (gal.):

8 fluid ounces = 1 cup
2 cups = 1 pint
2 pints = 1 quart
4 quarts = 1 gallon

Units of fluid volume are typically used for measuring the volume of things that can be poured. The volume of solid objects is more commonly measured in cubic units of distance, such as cubic inches and cubic feet.

- **Units of weight:** Weight is the measurement of how strongly gravity pulls an object toward Earth. Weight is measured in ounces (oz.), pounds (lb.), and tons.

  16 ounces = 1 pound
  2,000 pounds = 1 ton

Don't confuse *fluid ounces,* which measure volume, with *ounces,* which measure weight. These units are two completely different types of measurements!

- **Units of time:** Time is hard to define, but everybody knows what it is. Time is measured in seconds, minutes, hours, days, weeks, and years:

  60 seconds = 1 minute
  60 minutes = 1 hour
  24 hours = 1 day
  7 days = 1 week
  365 days ≈ 1 year

The conversion from days to years is approximate because Earth's daily rotation on its axis and its yearly revolution around the sun aren't exactly synchronized. A year is closer to 365.25 days, which is why leap-years exist.

I left months out of the picture because the definition of a month is imprecise — it can vary from 28 to 31 days.

- **Unit of speed:** Speed is the measurement of how much time an object takes to move a given distance. The most common unit of speed is miles per hour (mph).

- **Unit of temperature:** Temperature measures how much heat an object contains. This object can be a glass of water, a turkey in the oven, or the air surrounding your house. Temperature is measured in degrees Fahrenheit (°F).

Looking at the metric system

Like the English system, the metric system provides units of measurement for distance, volume, and so on. Unlike the English system, however, the metric system builds these units using a *basic unit* and a set of *prefixes.*

Table 15-1 shows five important basic units in the metric system.

Table 15-1 Five Basic Metric Units

Measure Of	*Basic Metric Unit*
Distance	Meter
Volume (capacity)	Liter
Mass (weight)	Gram
Time	Second
Temperature	Degrees Celsius (°C)

For scientific purposes, the metric system has been updated to the more rigorously defined *System of International Units (SI).* Each basic SI unit correlates directly to a measurable scientific process that defines it. In SI, the kilogram (not the gram) is the basic unit of mass, the kelvin is the basic unit of temperature, and the liter is not considered a basic unit. For technical reasons, scientists tend to use the more rigidly defined SI, but most other people use the looser metric system. In everyday practice, you can think of the units in Table 15-1 as basic units.

Table 15-2 shows ten metric prefixes, with the three most commonly used in bold and italicized (see Chapter 14 for more information on powers of ten).

Table 15-2 Ten Metric Prefixes

Prefix	*Meaning*	*Number*	*Power of Ten*
Giga-	One billion	1,000,000,000	10^9
Mega-	One million	1,000,000	10^6
Kilo-	***One thousand***	***1,000***	10^3

Prefix	***Meaning***	***Number***	***Power of Ten***
Hecta-	One hundred	100	10^2
Deca-	Ten	10	10^1
(none)	One	1	10^0
Deci-	One tenth	0.1	10^{-1}
Centi-	***One hundredth***	***0.01***	10^{-2}
Milli-	***One thousandth***	***0.001***	10^{-3}
Micro-	One millionth	0.000001	10^{-6}
Nano-	One billionth	0.000000001	10^{-9}

Large and small metric units are formed by linking a basic unit with a prefix. For example, linking the prefix *kilo-* to the basic unit *meter* gives you the *kilometer,* which means 1,000 meters. Similarly, linking the prefix *milli-* to the basic unit *liter* gives you the *milliliter,* which means 0.001 (one thousandth) of a meter.

Here's a list giving you the basics:

- **Units of distance:** The basic metric unit of distance is the meter (m). Other common units are millimeters (mm), centimeters (cm), and kilometers (km):

  $$1 \text{ kilometer} = 1{,}000 \text{ meters}$$
  $$1 \text{ meter} = 100 \text{ centimeters}$$
  $$1 \text{ meter} = 1{,}000 \text{ millimeters}$$

- **Units of fluid volume:** The basic metric unit of fluid volume (also called capacity) is the liter (L). Another common unit is the milliliter (mL):

  $$1 \text{ liter} = 1{,}000 \text{ milliliters}$$

  Note: One milliliter is equal to 1 cubic centimeter (cc).

- **Units of mass:** Technically, the metric system measures not weight, but mass. *Weight* is the measurement of how strongly gravity pulls an object toward Earth. *Mass,* however, is the measurement of the amount of matter an object has. If you traveled to the moon, your weight would change, so you would feel lighter. But your mass would remain the same, so all of you would still be there. Unless you're planning a trip into outer space or performing a scientific experiment, you probably don't need to know the difference between weight and mass. In this chapter, you can think of them as equivalent, and I use the word *weight* when referring to metric mass.

The basic unit of weight in the metric system is the gram (g). Even more commonly used, however, is the kilogram (kg):

1 kilogram = 1,000 grams

Note: 1 kilogram of water has a volume of 1 liter.

- **Units of time:** As in the English system, the basic metric unit of time is a second (s). For most purposes, people also use other English units, such as minutes and hours.

  For many scientific purposes, the second is the only unit used to measure time. Large numbers of seconds and small fractions of sections are represented with *scientific notation,* which I cover in Chapter 14.

- **Units of speed:** For most purposes, the most common metric unit of speed (also called velocity) is kilometers per hour (km/hr). Another common unit is *meters per second* (m/s).

- **Units of temperature (degrees Celsius or Centigrade):** The basic metric unit of temperature is the Celsius degree (°C), also called the *Centigrade degree.* The Celsius scale is set up so that, at sea level, water freezes at 0°C and boils at 100°C.

  Scientists often use another unit — the kelvin (K) — to talk about temperature. The degrees are the same size as in Celsius, but 0 K is set at *absolute zero,* the temperature at which atoms don't move at all. Absolute zero is approximately equal to –273.15°C.

Estimating and Converting between the English and Metric Systems

Most Americans use the English system of measurement all the time and have only a passing acquaintance with the metric system. But metric units are being used more commonly as the units for tools, footraces, soft drinks, and many other things. Also, if you travel abroad, you need to know how far 100 kilometers is or how long you can drive on 10 liters of gasoline.

In this section, I show you how to make ballpark estimates of metric units in terms of English units, which can help you feel more comfortable with metric units. I also show you how to convert between English and metric units, which is a common type of math problem.

When I talk about *estimating,* I mean very loose ways of measuring metric amounts using the English units you are familiar with. In contrast, when I talk about *converting,* I mean using an equation to change from one system of units to the other. Neither method is exact, but converting provides a much closer approximation (and takes longer) than estimating.

Estimating in the metric system

One reason people sometimes feel uncomfortable using the metric system is that, when you're not familiar with it, estimating amounts in practical terms is hard. For example, if I tell you that we're going out to a beach that's $\frac{1}{4}$ mile away, you prepare yourself for a short walk. And if I tell you that it's 10 miles away, you head for the car. But what do you do with the information that the beach is 3 kilometers away?

Similarly, if I tell you that the temperature is 85°F, you'll probably wear a bathing suit or shorts. And if I tell you it's 40°F, you'll probably wear a coat. But what do you wear if I tell you that the temperature is 25°C?

In this section, I give you a few rules of thumb to estimate metric amounts. In each case, I show you how a common metric unit compares with an English unit that you already feel comfortable with.

Approximating short distances: 1 meter is about 1 yard (3 feet)

Here's how to convert meters to feet: 1 meter ≈ 3.28 feet. But for estimating, use the simple rule that 1 meter is about 1 yard (that is, about 3 feet).

By this estimate, a 6-foot man stands about 2 meters tall. A 15-foot room is 5 meters wide. And a football field that's 100 yards long is about 100 meters long. Similarly, a river with a depth of 4 meters is about 12 feet deep. A mountain that's 3,000 meters tall is about 9,000 feet. And a child who is only half a meter tall is about a foot and a half.

Estimating longer distances and speed

Here's how to convert kilometers to miles: 1 kilometer ≈ 0.62 miles. For a ballpark estimate, you can remember that 1 kilometer is about $\frac{1}{2}$ a mile. By the same token, 1 kilometer per hour is about $\frac{1}{2}$ mile per hour.

This guideline tells you that if you live 2 miles from the nearest supermarket, then you live about 4 kilometers from there. A marathon of 26 miles is about 52 kilometers. And if you run on a treadmill at 6 miles per hour, then you can run at about 12 kilometers per hour. By the same token, a 10-kilometer race is about 5 miles. If the Tour de France is about 4,000 kilometers, then it's about

2,000 miles. And if light travels about 300,000 kilometers per second, then it travels about 150,000 miles per second.

Approximating volume: 1 liter is about 1 quart (1/4 gallon)

Here's how to convert liters to gallons: 1 liter ≈ 0.26 gallons. A good estimate here is that 1 liter is about 1 quart (a gallon consists of about 4 liters).

Using this estimate, a gallon of milk is 4 quarts, so it's about 4 liters. If you put 10 gallons of gasoline in your tank, it's about 40 liters. In the other direction, if you buy a 2-liter bottle of cola, you have about 2 quarts. If you buy an aquarium with a 100-liter capacity, it holds about 25 gallons of water. And if a pool holds 8,000 liters of water, it holds about 2,000 gallons.

Estimating weight: 1 kilogram is about 2 pounds

Here's how to convert kilograms to pounds: 1 kilogram ≈ 2.20 pounds. For estimating, figure that 1 kilogram is equal to about 2 pounds.

By this estimate, a 5-kilogram bag of potatoes weighs about 10 pounds. If you can bench-press 70 kilograms, then you can bench-press about 140 pounds. And because a liter of water weighs exactly 1 kilogram, you know that a quart of water weighs about 2 pounds. Similarly, if a baby weighs 8 pounds at birth, he or she weighs about 4 kilograms. If you weigh 150 pounds, then you weigh about 75 kilograms. And if your New Year's resolution is to lose 20 pounds, then you want to lose about 10 kilograms.

Estimating temperature

The most common reason for estimating temperature in Celsius is in connection with the weather. The formula for converting from Celsius to Fahrenheit is kind of messy:

$$\text{Fahrenheit} = \text{Celsius} \times \frac{9}{5} + 32$$

Instead, use the handy chart in Table 15-3.

Table 15-3 Comparing Celsius and Fahrenheit Temperatures

Celsius (Centigrade)	*Description*	*Fahrenheit*
0°	Cold	32°
10°	Cool	50°
20°	Warm	68°
30°	Hot	86°

Any temperature below 0°C is cold, and any temperature over 30°C is hot. Most of the time, the temperature falls in this middling range. So now you know that when the temperature is 6°C, you want to wear a coat. When it's 14°C, you may want a sweater — or at least long sleeves. And when it's 25°C, head for the beach!

Converting units of measurement

Many books give you one formula for converting from English to metric and another for converting from metric to English. People often find this conversion method confusing because they have trouble remembering which formula to use in which direction.

In this section, I show you a simple way to convert between English and metric units that uses only one formula for each type of conversion.

Here's a nice pair that's easy to remember: 16°C is about 61°F.

Understanding conversion factors

When you multiply any number by 1, that number stays the same. For example, 36 × 1 = 36. And when a fraction has the same numerator (top number) and denominator (bottom number), that fraction equals 1 (see Chapter 10 for details). So when you multiply a number by a fraction that equals 1, the number stays the same. For example:

$$36 \times \frac{5}{5} = 36$$

If you multiply a measurement by a special fraction that equals 1, you can switch from one unit of measurement to another without changing the value. People call such fractions *conversion factors.*

Take a look at some equations that show how metric and English units are related (all conversions between English and metric units are approximate):

- 1 meter ≈ 3.26 feet
- 1 kilometer ≈ 0.62 mile
- 1 liter ≈ 0.26 gallon
- 1 kilogram ≈ 2.20 pounds

Because the values on each side of the equations are equal, you can create

- $\frac{1\text{ meter}}{3.26\text{ feet}}$ or $\frac{3.26\text{ feet}}{1\text{ meter}}$
- $\frac{1\text{ kilometer}}{0.62\text{ mile}}$ or $\frac{0.62\text{ mile}}{1\text{ kilometer}}$
- $\frac{1\text{ liter}}{0.26\text{ gallon}}$ or $\frac{0.26\text{ gallon}}{1\text{ liter}}$
- $\frac{1\text{ kilogram}}{2.2\text{ pounds}}$ or $\frac{2.2\text{ pounds}}{1\text{ kilogram}}$

When you understand how units of measurement cancel (which I discuss in the next section), you can easily choose which fractions to use to switch between units of measurement.

Canceling units of measurement

When you're multiplying fractions, you can cancel any factor that appears in both the numerator and the denominator (see Chapter 9 for details). Just as with numbers, you can also cancel out units of measurement in fractions. For example, suppose you want to evaluate this fraction:

$$\frac{6\text{ gallons}}{2\text{ gallons}}$$

You already know that you can cancel out a factor of 2 in both the numerator and the denominator. But you can also cancel out the unit *gallons* in both the numerator and the denominator:

$$=\frac{6\text{ gallons}}{2\text{ gallons}}$$

So this fraction simplifies as follows:

$$=3$$

Converting units

When you understand how to cancel out units in fractions and how to set up fractions equal to 1 (see the preceding sections), you have a foolproof system for converting units of measurement.

Suppose you want to convert 7 meters into feet. Using the equation 1 meter = 3.26 feet, you can make a fraction out of the two values, as follows:

$$\frac{1\text{ meter}}{3.26\text{ feet}} = 1 \quad \text{or} \quad \frac{3.26\text{ feet}}{1\text{ meter}} = 1$$

Both fractions equal 1 because the numerator and the denominator are equal. So you can multiply the quantity you're trying to convert (7 meters) by one of these fractions without changing it. Remember that you want the meters unit to cancel out. You already have the word *meters* in the numerator (to make this clear, place 1 in the denominator), so use the fraction that puts *1 meter* in the denominator:

$$\frac{7\text{ meters}}{1} \times \frac{3.26\text{ feet}}{1\text{ meter}}$$

Now cancel out the unit that appears in both the numerator and the denominator:

$$= \frac{7\text{ }\cancel{\text{meters}}}{1} \times \frac{3.26\text{ feet}}{1\text{ }\cancel{\text{meter}}}$$

At this point, the only value in the denominator is 1, so you can ignore it. And the only unit left is *feet,* so place it at the end of the expression:

$$= 7 \times 3.26\text{ feet}$$

Now do the multiplication (Chapter 11 shows how to multiply decimals):

$$= 22.82\text{ feet}$$

It may seem strange that the answer appears with the units already attached, but that's the beauty of this method: When you set up the right expression, the answer just appears.

You can get more practice converting units of measurement in Chapter 18, where I show you how to set up conversion chains and tackle word problems involving measurement.

Chapter 16

Picture This: Basic Geometry

In This Chapter

- Knowing the basic components of geometry: points, lines, angles, and shapes
- Examining two-dimensional shapes
- Looking at solid geometry
- Finding out how to measure a variety of shapes

Geometry is the mathematics of figures such as squares, circles, triangles, and lines. Because geometry is the math of physical space, it's one of the most useful areas of math. Geometry comes into play when measuring rooms or walls in your house, the area of a circular garden, the volume of water in a pool, or the shortest distance across a rectangular field.

Although geometry is usually a yearlong course in high school, you may be surprised by how quickly you can pick up what you need to know about basic geometry. Much of what you discover in a geometry course is how to write geometric proofs, which you don't need for algebra — or trigonometry, or even calculus.

In this chapter, I give you a quick and practical overview of geometry. First, I show you four important concepts in plane geometry: points, lines, angles, and shapes. Then I give you the basics on geometric shapes, from flat circles to solid cubes. Finally, I discuss how to measure geometric shapes by finding the area and perimeter of two-dimensional forms and the volume and surface area of some geometric solids.

Of course, if you want to know more about geometry, the ideal place to look beyond this chapter is *Geometry For Dummies*, 2nd Edition, by Mark Ryan (published by Wiley)!

Getting on the Plane: Points, Lines, Angles, and Shapes

Plane geometry is the study of figures on a two-dimensional surface — that is, on a *plane*. You can think of the plane as a piece of paper with no thickness at all. Technically, a plane doesn't end at the edge of the paper — it continues forever.

In this section, I introduce you to four important concepts in plane geometry: points, lines, angles, and shapes (squares, circles, triangles, and so forth).

Making some points

A *point* is a location on a plane. It has no size or shape. Although in reality a point is too small to be seen, you can represent it visually in a drawing by using a dot.

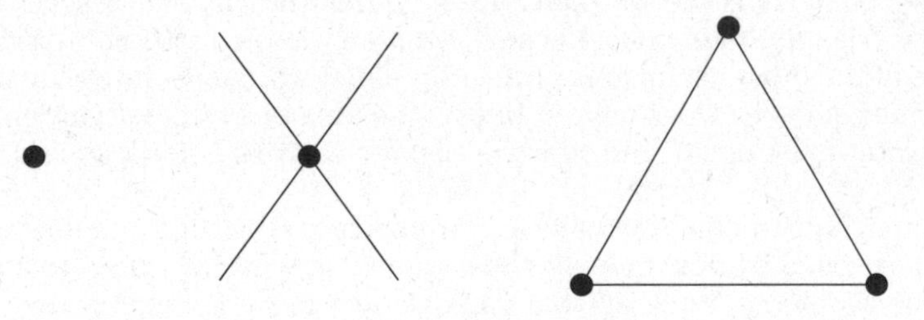

Illustration by Wiley, Composition Services Graphics

When two lines intersect, as shown in this figure, they share a single point. Additionally, each corner of a polygon is a point. (Keep reading for more on lines and polygons.)

Knowing your lines

A *line* — also called a *straight line* — is pretty much what it sounds like; it marks the shortest distance between two points, but it extends infinitely in both directions. It has length but no width, making it a one-dimensional (1-D) figure.

Given any two points, you can draw exactly one line that passes through both of them. In other words, two points *determine* a line.

Illustration by Wiley, Composition Services Graphics

When two lines intersect, they share a single point. When two lines don't intersect, they are *parallel,* which means that they remain the same distance from each other everywhere. A good visual aid for parallel lines is a set of railroad tracks. In geometry, you draw a line with arrows at both ends. Arrows on either end of a line mean that the line goes on forever (as you can see in Chapter 1, where I discuss the number line).

A *line segment* is a piece of a line that has endpoints, as shown here.

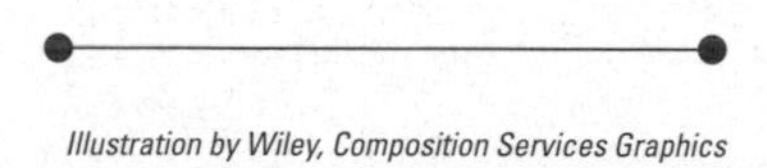

Illustration by Wiley, Composition Services Graphics

A *ray* is a piece of a line that starts at a point and extends infinitely in one direction, kind of like a laser. It has one endpoint and one arrow.

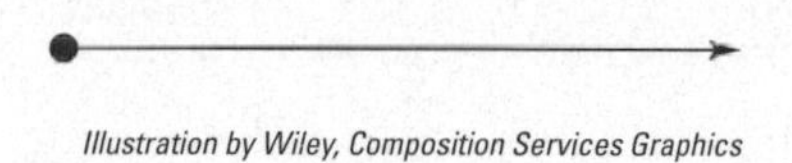

Illustration by Wiley, Composition Services Graphics

Figuring the angles

An *angle* is formed when two rays extend from the same point.

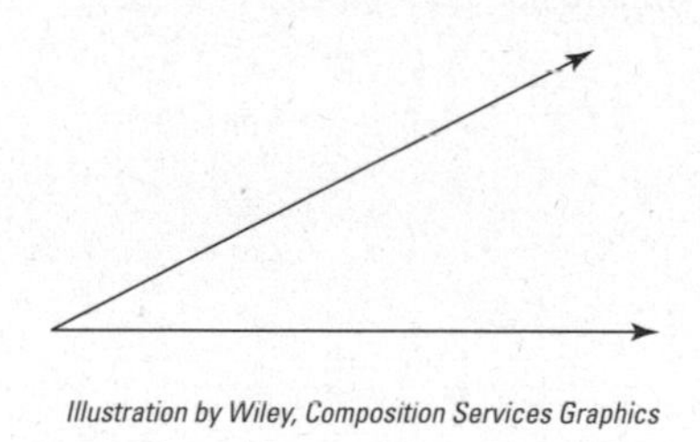

Illustration by Wiley, Composition Services Graphics

Angles are typically used in carpentry to measure the corners of objects. They're also used in navigation to indicate a sudden change in direction. For example, when you're driving, it's common to distinguish when the angle of a turn is "sharp" or "not so sharp."

The sharpness of an angle is usually measured in *degrees.* The most common angle is the *right angle* — the angle at the corner of a square — which is a 90° (90-degree) angle:

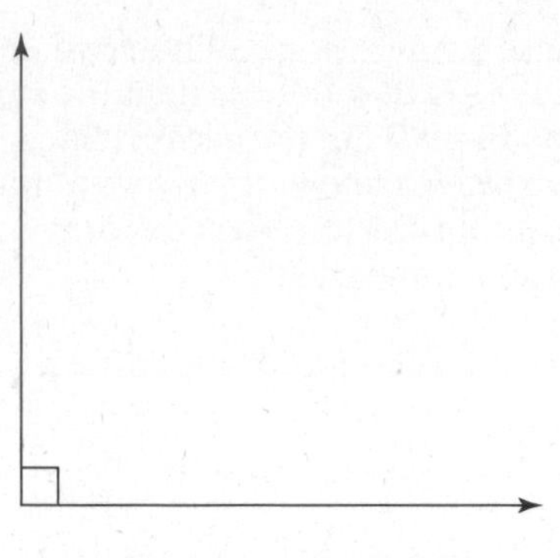

Illustration by Wiley, Composition Services Graphics

Angles that have fewer than 90° — that is, angles that are sharper than a right angle — are called *acute angles,* like this one:

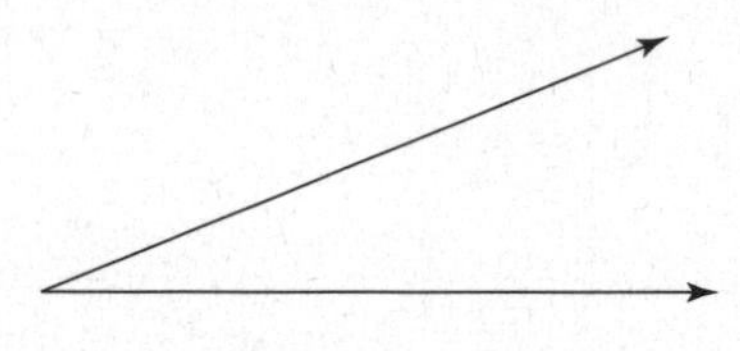

Illustration by Wiley, Composition Services Graphics

Angles that measure greater than 90° — that is, angles that aren't as sharp as a right angle — are called *obtuse angles,* as seen here:

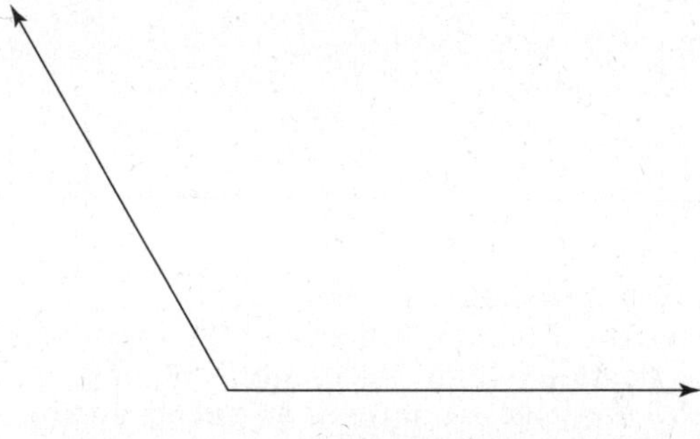

Illustration by Wiley, Composition Services Graphics

When an angle is exactly 180°, it forms a straight line and is called a *straight angle.*

Illustration by Wiley, Composition Services Graphics

Shaping things up

A shape is any closed geometrical figure that has an inside and an outside. Circles, squares, triangles, and larger polygons are all examples of shapes.

Much of plane geometry focuses on different types of shapes. In the next section, I show you how to identify a variety of shapes. Later in this chapter, I show you how to measure these shapes.

Closed Encounters: Shaping Up Your Understanding of 2-D Shapes

A *shape* is any closed two-dimensional (2-D) geometrical figure that has an inside and an *outside*, separated by the *perimeter* (boundary) of the shape. The *area* of a shape is the measurement of the size inside that shape.

A few shapes that you're probably familiar with include the square, rectangle, and triangle. However, many shapes don't have names, as you can see in Figure 16-1.

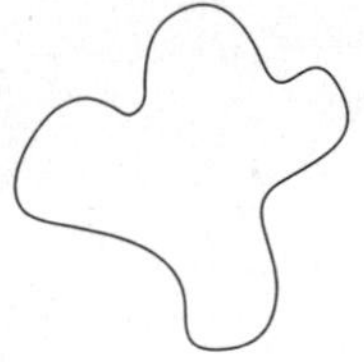
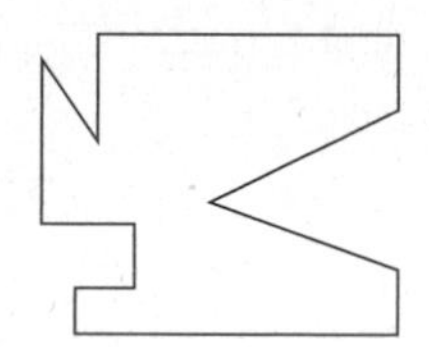

Figure 16-1: Unnamed shapes.

Illustration by Wiley, Composition Services Graphics

Measuring the perimeter and area of shapes is useful for a variety of applications, from land surveying (to get information about a parcel of land that you're measuring) to sewing (to figure out how much material you need for a project). In this section, I introduce you to a variety of geometric shapes. Later in the chapter, I show you how to find the perimeter and area of each, but for now, I just acquaint you with them.

Polygons

A *polygon* is any shape whose sides are all straight. Every polygon has three or more sides (if it had fewer than three, it wouldn't really be a shape at all). Following are a few of the most common polygons.

Triangles

The most basic shape with straight sides is the *triangle,* a three-sided polygon. You find out all about triangles when you study trigonometry (and what better place to begin than *Trigonometry For Dummies*, 2nd Edition, by Mary Jane Sterling [Wiley]?). Triangles are classified on the basis of their sides and angles. Take a look at the differences (and see Figure 16-2):

- **Equilateral:** An *equilateral triangle* has three sides that are all the same length and three angles that all measure 60°.
- **Isosceles:** An *isosceles triangle* has two sides that are the same length and two equal angles.
- **Scalene:** *Scalene triangles* have three sides that are all different lengths and three angles that are all unequal.
- **Right:** A *right triangle* has one right angle. It may be isosceles or scalene.

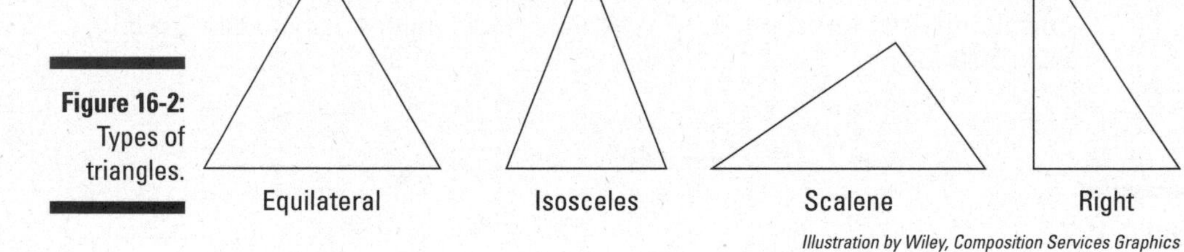

Figure 16-2: Types of triangles.

Illustration by Wiley, Composition Services Graphics

Quadrilaterals

A *quadrilateral* is any shape that has four straight sides. Quadrilaterals are one of the most common shapes you see in daily life. If you doubt this statement, look around and notice that most rooms, doors, windows, and tabletops are quadrilaterals. Here I introduce you to a few common quadrilaterals (Figure 16-3 shows you what they look like):

- **Square:** A *square* has four right angles and four sides of equal length; also, both pairs of opposite sides (sides directly across from each other) are parallel.

- **Rectangle:** Like a square, a *rectangle* has four right angles and two pairs of opposite sides that are parallel. Unlike the square, however, although opposite sides are equal in length, sides that share a corner — *adjacent* sides — may have different lengths.
- **Rhombus:** Imagine starting with a square and collapsing it as if its corners were hinges. This shape is called a *rhombus*. All four sides are equal in length, and both pairs of opposite sides are parallel.
- **Parallelogram:** Imagine starting with a rectangle and collapsing it as if the corners were hinges. This shape is a *parallelogram* — both pairs of opposite sides are equal in length, and both pairs of opposite sides are parallel.
- **Trapezoid:** The *trapezoid*'s only important feature is that at least two opposite sides are parallel.
- **Kite:** A kite is a quadrilateral with two pairs of adjacent sides that are the same length.

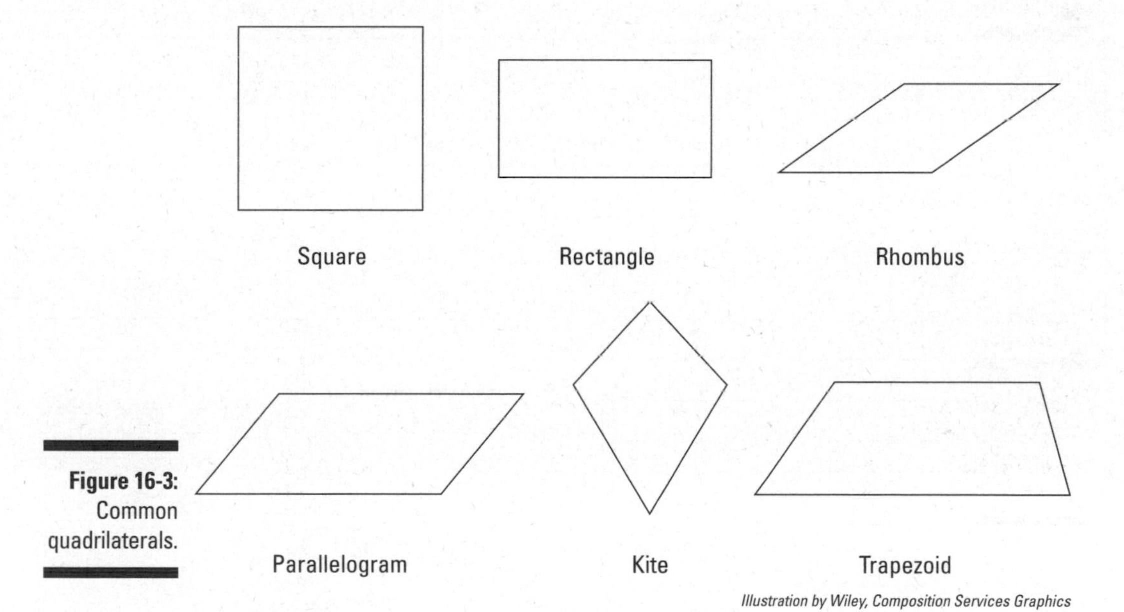

Figure 16-3: Common quadrilaterals.

Illustration by Wiley, Composition Services Graphics

A quadrilateral can fit into more than one of these categories. For example, every parallelogram (with two sets of parallel sides) is also a trapezoid (with at least one set of parallel sides). Every rectangle and rhombus is also both a parallelogram and a trapezoid. And every square is also all five other types of quadrilaterals. In practice, however, it's common to identify a quadrilateral as descriptively as possible — that is, use the *first* word in the list that accurately describes it.

Polygons on steroids — larger polygons

A polygon can have any number of sides. Polygons with more than four sides aren't as common as triangles and quadrilaterals, but they're still worth knowing about. Larger polygons come in two basic varieties: regular and irregular.

A *regular polygon* has equal sides and equal angles. The most common are regular pentagons (five sides), regular hexagons (six sides), and regular octagons (eight sides). See Figure 16-4.

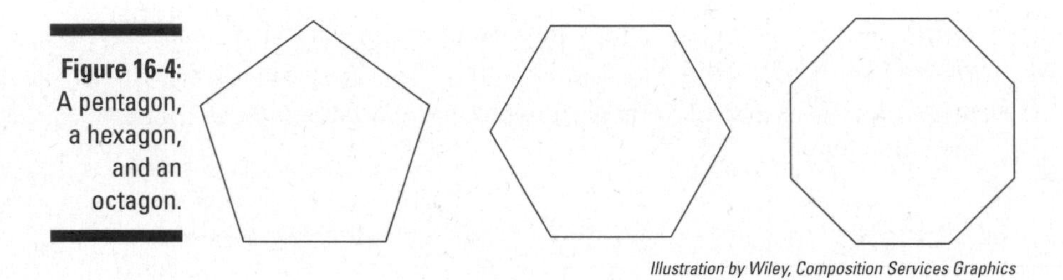

Figure 16-4: A pentagon, a hexagon, and an octagon.

Illustration by Wiley, Composition Services Graphics

Every other polygon is an *irregular polygon* (see Figure 16-5).

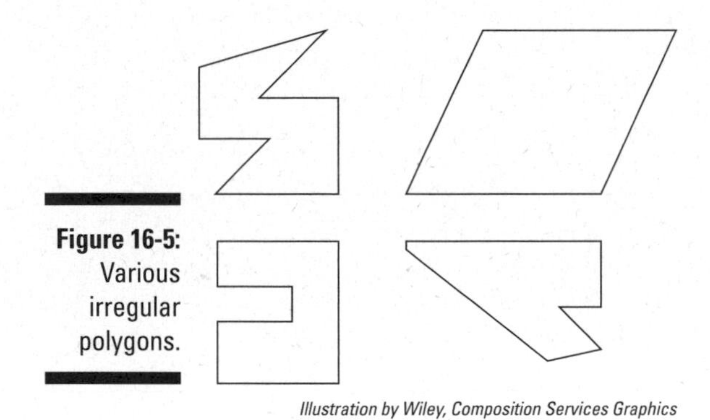

Figure 16-5: Various irregular polygons.

Illustration by Wiley, Composition Services Graphics

Circles

A circle is the set of all points that are a constant distance from the circle's center. The distance from any point on the circle to its center is called the *radius* of the circle. The distance from any point on the circle straight through the center to the other side of the circle is called the *diameter* of the circle.

Unlike polygons, a circle has no straight edges. The ancient Greeks — who invented much of geometry as we know it today — thought that the circle was the most perfect geometric shape.

Taking a Trip to Another Dimension: Solid Geometry

Solid geometry is the study of shapes in *space* — that is, the study of shapes in three dimensions. A *solid* is the spatial (three-dimensional, or 3-D) equivalent of a shape. Every solid has an *inside* and an *outside* separated by the surface of the solid. Here, I introduce you to a variety of solids.

The many faces of polyhedrons

A *polyhedron* is the three-dimensional equivalent of a polygon. As you may recall from earlier in the chapter, a polygon is a shape that has only straight sides. Similarly, a polyhedron is a solid that has only straight edges and flat faces (that is, faces that are polygons).

The most common polyhedron is the *cube* (see Figure 16-6). As you can see, a cube has 6 flat faces that are polygons — in this case, all the faces are square — and 12 straight edges. Additionally, a cube has eight *vertexes*, or *vertices* (corners). Later in this chapter, I show you how to measure the surface area and volume of a cube.

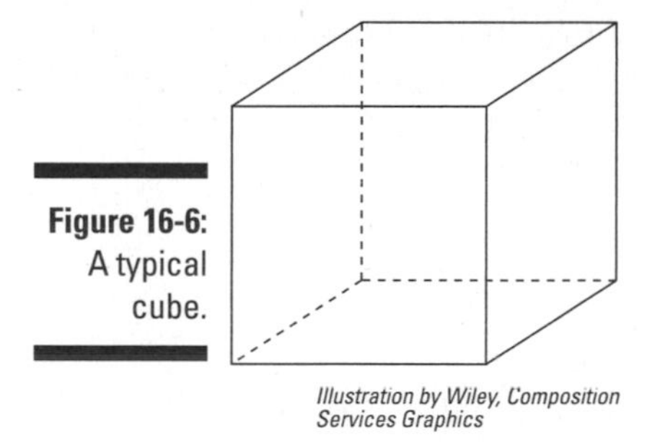

Figure 16-6: A typical cube.

Illustration by Wiley, Composition Services Graphics

Figure 16-7 shows a few common polyhedrons (or polyhedra).

Later in this chapter, I show you how measure each of these polyhedrons to determine its volume — that is, the amount of space contained inside its surface.

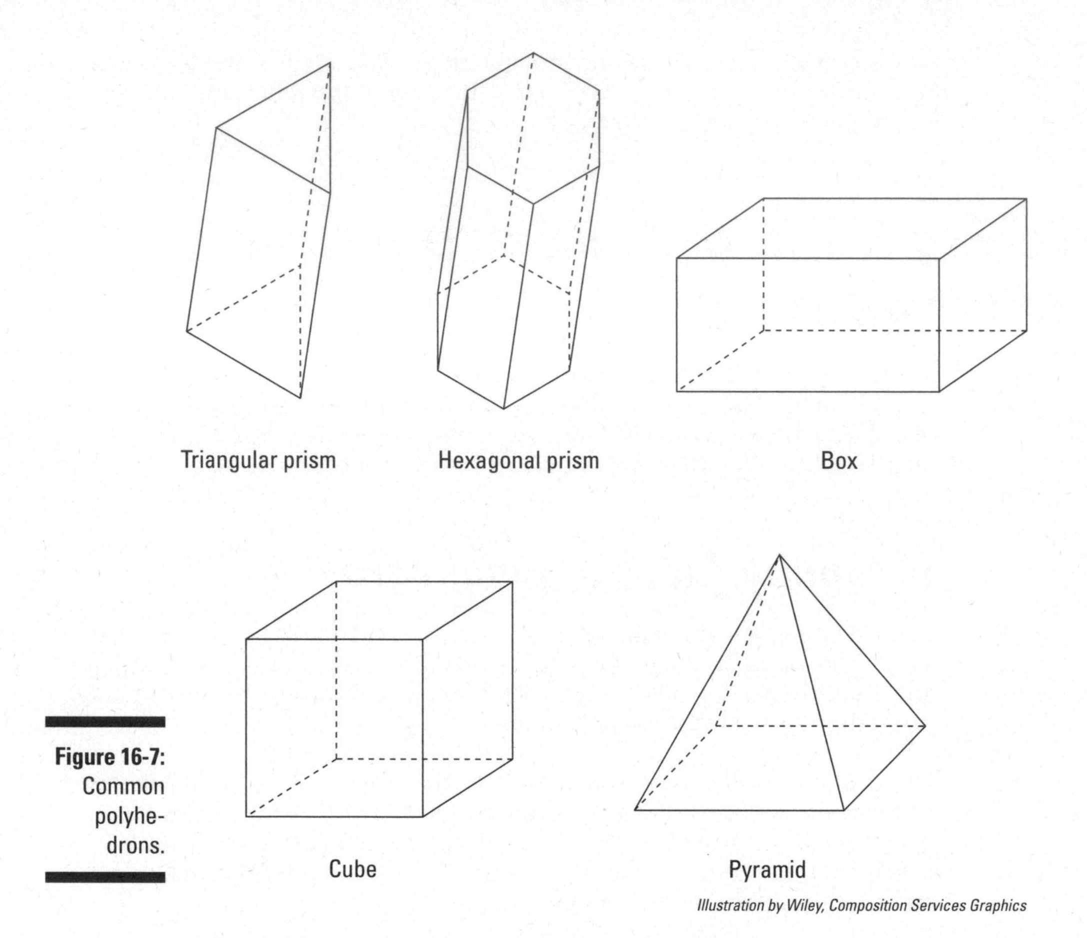

Figure 16-7: Common polyhedrons.

Illustration by Wiley, Composition Services Graphics

One special set of polyhedrons is called the *five regular solids* (see Figure 16-8). Each regular solid has identical faces that are regular polygons. Notice that a cube is a type of regular solid. Similarly, the tetrahedron is a pyramid with four faces that are equilateral triangles.

3-D shapes with curves

Many solids aren't polyhedrons because they contain at least one curved surface. Here are a few of the most common of these types of solids (also see Figure 16-9):

- **Sphere:** A *sphere* is the solid, or three-dimensional, equivalent of a circle. A ball is a perfect visual aid for a sphere.

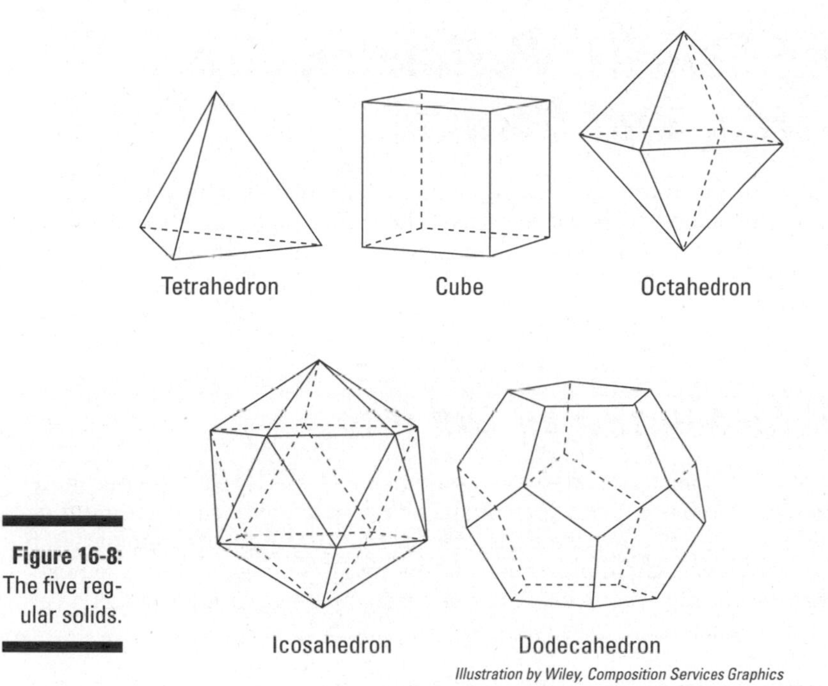

Figure 16-8: The five regular solids.

Illustration by Wiley, Composition Services Graphics

- **Cylinder:** A *cylinder* has a circular base and extends vertically from the plane. A good visual aid for a cylinder is a can of soup.
- **Cone:** A *cone* is a solid with a round base that extends vertically to a single point. A good visual aid for a cone is an ice-cream cone.

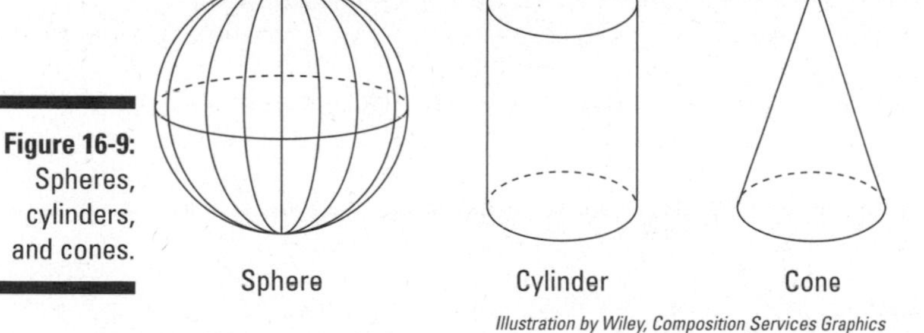

Figure 16-9: Spheres, cylinders, and cones.

Illustration by Wiley, Composition Services Graphics

In the next section, I show you how to measure a sphere and a cylinder to determine their volume — that is, the amount of space contained within.

Measuring Shapes: Perimeter, Area, Surface Area, and Volume

In this section, I introduce you to some important formulas for measuring shapes on the plane and solids in space. These formulas use letters to stand for numbers that you can plug in to make specific measurements. Using letters in place of numbers is a feature you'll see more of in Part V, when I discuss algebra.

2-D: Measuring on the flat

Two important skills in geometry — and real life — are finding the perimeter and calculating the area of shapes. A shape's *perimeter* is a measurement of the length of its sides. You use perimeter for measuring the distance around the edges of a room, building, or circular pathway. A shape's *area* is a measurement of how big it is inside. You use area when measuring the size of a wall, a table, or a tire.

For example, in Figure 16-10, I give you the lengths of the sides of each shape.

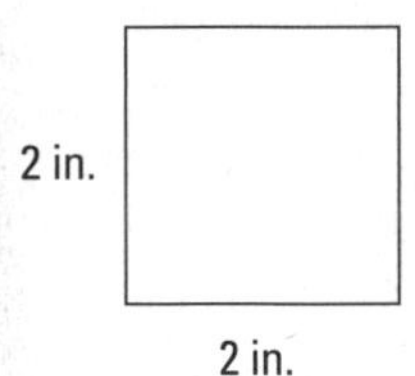

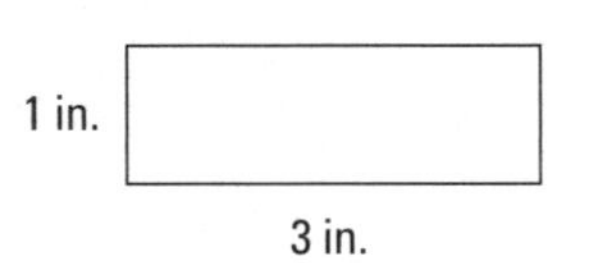

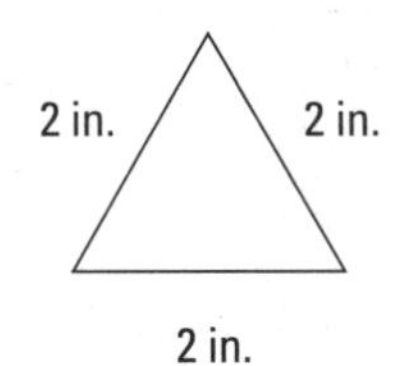

Figure 16-10: Measuring the sides of figures.

Illustration by Wiley, Composition Services Graphics

When every side of a shape is straight, you can measure its perimeter by adding up the lengths of all its sides.

Similarly, in Figure 16-11, I give you the area of each shape.

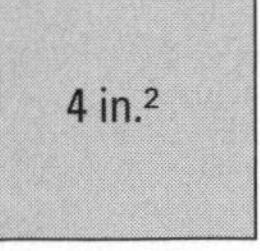

3 in.²

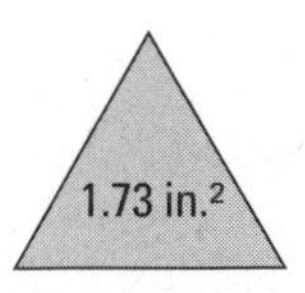

Figure 16-11: The areas of figures.

Illustration by Wiley, Composition Services Graphics

The area of a shape is always measured in *square units:* square inches (in.^{2}), square feet (ft.^{2}), square miles (mi.^{2}), square kilometers (km^2), and so on — even if you're talking about the area of a circle! (For more on measurements, flip to Chapter 15.)

I cover these types of calculations in this section. (For more information on the names of shapes, refer to "Closed Encounters: Shaping Up Your Understanding of 2-D Shapes.")

Measuring squares

The letter s represents the length of a square's side. For example, if the side of a square is 3 inches, then you say $s = 3$ in. Finding the perimeter *(P)* of a square is simple: Just multiply the length of the side by 4. Here's the formula for the perimeter of a square:

$$P = 4 \times s$$

For example, if the length of the side is 3 inches, substitute 3 inches for s in the formula:

$$P = 4 \times 3\,\text{in.} = 12\,\text{in.}$$

Finding the area (A) of a square is also easy: Just multiply the length of the side by itself — that is, take the *square* of the side. Here are two ways of writing the formula for the area of a square (s^2 is pronounced "s squared"):

$$A = s^2 \quad \text{or} \quad A = s \times s$$

For example, if the length of the side is 3 inches, then you get the following:

$$A = (3\,\text{in.})^2 = 3\text{ in.} \times 3\text{ in.} = 9\,\text{in.}^2$$

Working with rectangles

The long side of a rectangle is called the *length,* or l for short. The short side is called the *width,* or w for short. For example, in a rectangle whose sides are 5 and 4 feet long, $l = 5$ ft. and $w = 4$ ft.

Because a rectangle has two lengths and two widths, you can use the following formula for the perimeter of a rectangle:

$$P = 2 \times (l + w)$$

Calculate the perimeter of a rectangle whose length is 5 yards and whose width is 4 yards as follows:

$$P = 2\times(5\text{ yd.} + 4\text{ yd.}) = 2\times 9\text{ yd.} = 18\text{ yd.}$$

The formula for the area of a rectangle is:

$$A = l\times w$$

So here's how you calculate the area of the same rectangle:

$$A = l\times w = 5\text{ yd.}\times 4\text{ yd.} = 20\text{ yd.}^2$$

Calculating with rhombuses

As with a square, use *s* to represent the length of a rhombus's side. But another key measurement for a rhombus is its height. The *height* of a rhombus (*h* for short) is the shortest distance from one side to the opposite side. In Figure 16-12, $s = 4$ cm and $h = 2$ cm.

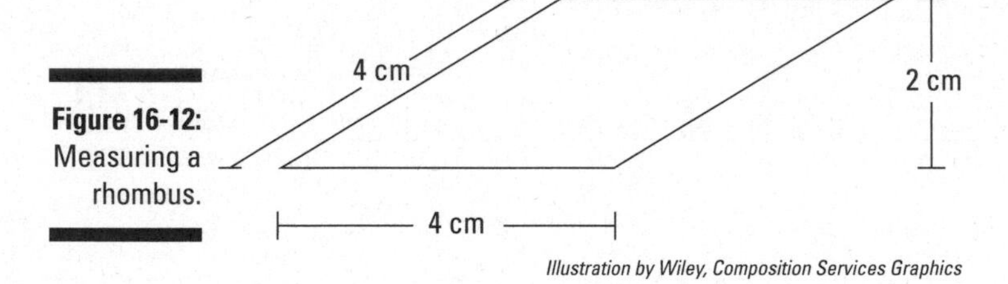

Figure 16-12: Measuring a rhombus.

Illustration by Wiley, Composition Services Graphics

The formula for the perimeter of a rhombus is the same as for a square:

$$P = 4\times s$$

Here's how you figure out the perimeter of a rhombus whose side is 4 centimeters:

$$P = 4\times 4\text{ cm} = 16\text{ cm}$$

To measure the area of a rhombus, you need both the length of the side and the height. Here's the formula:

$$A = s\times h$$

So here's how you determine the area of a rhombus with a side of 4 cm and a height of 2 cm:

$$A = 4\,\text{cm} \times 2\,\text{cm} = 8\,\text{cm}^2$$

You can read 8 cm^2 as "8 square centimeters" or, less commonly, as "8 centimeters squared."

Measuring parallelograms

The top and bottom sides of a parallelogram are called its *bases* (b for short), and the remaining two sides are its *sides* (s). And as with rhombuses, another important measurement of a parallelogram is its *height* (h), the shortest distance between the bases. So the parallelogram in Figure 16-13 has these measurements: $b = 6$ in., $s = 3$ in., and $h = 2$ in.

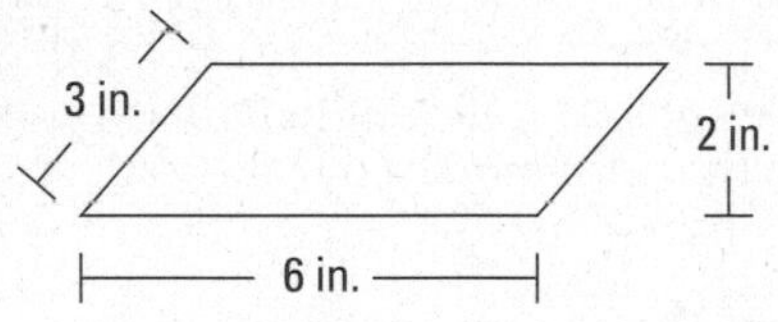

Figure 16-13: Measuring a parallelogram.

Illustration by Wiley, Composition Services Graphics

Each parallelogram has two equal bases and two equal sides. Therefore, here's the formula for the perimeter of a parallelogram:

$$P = 2 \times (b + s)$$

To figure out the perimeter of the parallelogram in this section, just substitute the measurements for the bases and sides:

$$P = 2(6\,\text{in.} + 3\,\text{in.}) = 2 \times 9\,\text{in.} = 18\,\text{in.}$$

And here's the formula for the area of a parallelogram:

$$A = b \times h$$

Here's how you calculate the area of the same parallelogram:

$$A = 6\,\text{in.} \times 2\,\text{in.} = 12\,\text{in.}^2$$

Measuring trapezoids

The parallel sides of a trapezoid are called its *bases*. Because these bases are different lengths, you can call them *b1* and *b2*. The height *(h)* of a trapezoid is the shortest distance between the bases. Thus, the trapezoid in Figure 16-14 has these measurements: $b1 = 2$ in., $b2 = 3$ in., and $h = 2$ in.

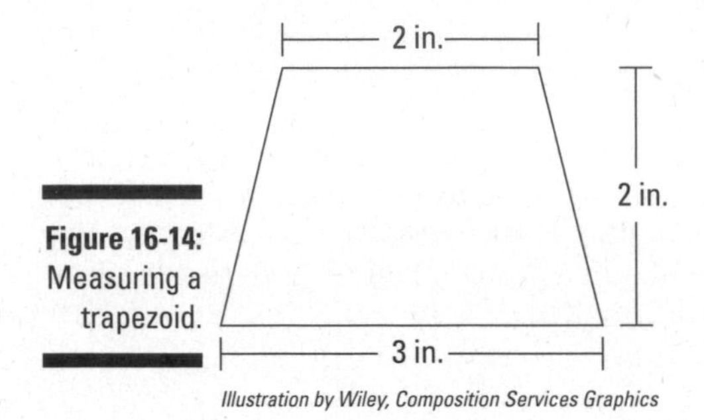

Figure 16-14: Measuring a trapezoid.

Illustration by Wiley, Composition Services Graphics

Because a trapezoid can have sides of four different lengths, you really don't have a special formula for finding the perimeter of a trapezoid. Just add up the lengths of its sides, and you get your answer.

Here's the formula for the area of a trapezoid:

$$A = \frac{1}{2} \times (b_1 + b_2) \times h$$

So here's how to find the area of the pictured trapezoid:

$$\begin{aligned} A &= \frac{1}{2} \times (2\,\text{in.} + 3\,\text{in.}) \times 2\,\text{in.} \\ &= \frac{1}{2} \times 5\,\text{in.} \times 2\,\text{in.} \\ &= 5\,\text{in.}^2 \end{aligned}$$

Measuring triangles

In this section, I discuss how to measure the perimeter and area of all triangles. Then I show you a special feature of right triangles that allows you to measure them more easily.

Finding the perimeter and area of a triangle

Mathematicians have no special formula for finding the perimeter of a triangle — they just add up the lengths of the sides.

To find the area of a triangle, you need to know the length of one side — the base (*b* for short) — and the height (*h*). Note that the height forms a right angle with the base. Figure 16-15 shows a triangle with a base of 5 cm and a height of 2 cm:

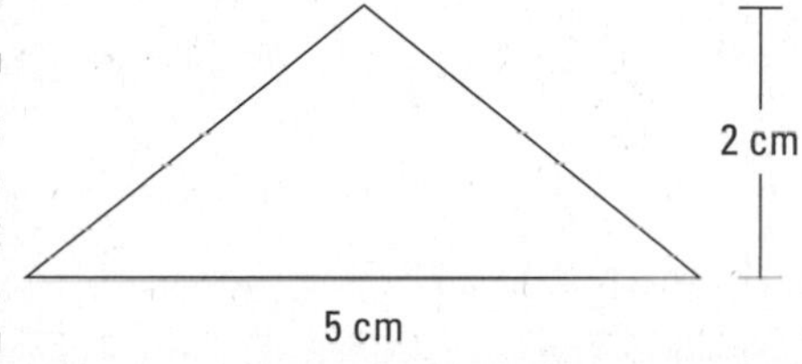

Figure 16-15: The base and height of a triangle.

Illustration by Wiley, Composition Services Graphics

Here's the formula for the area of a triangle:

$$A = \frac{1}{2} \times b \times h$$

So here's how to figure out the area of a triangle with a base of 5 cm and a height of 2 cm:

$$A = \frac{1}{2} \times 5\,\text{cm} \times 2\,\text{cm} = \frac{1}{2} \times 10\,\text{cm}^2 = 5\,\text{cm}^2$$

Lessons from Pythagoras: Finding the third side of a right triangle

The long side of a right triangle (*c*) is called the *hypotenuse,* and the two short sides (*a* and *b*) are called the *legs* (see Figure 16-16). The most important right triangle formula is the *Pythagorean theorem*:

$$a^2 + b^2 = c^2$$

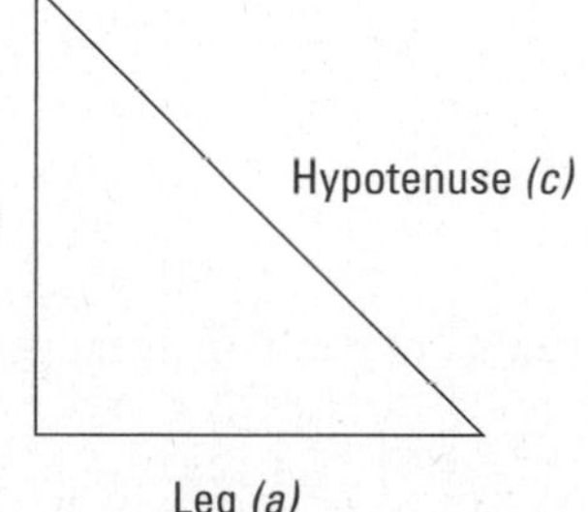

Figure 16-16: The hypotenuse and legs of a right triangle.

Illustration by Wiley, Composition Services Graphics

This formula allows you to find the hypotenuse of a triangle, given only the lengths of the legs. For example, suppose the legs of a triangle are 3 and 4 units. Here's how to use the Pythagorean theorem to find the length of the hypotenuse:

$$3^2 + 4^2 = c^2$$
$$9 + 16 = c^2$$
$$25 = c^2$$

So when you multiply c by itself, the result is 25. Therefore,

$$c = 5$$

The length of the hypotenuse is 5 units.

Going 'round in circles

The *center* of a circle is a point that's the same distance from any point on the circle itself. This distance is called the *radius* of the circle, or r for short. And any line segment from one point on the circle through the center to another point on the circle is called a *diameter,* or d for short. See Figure 16-17.

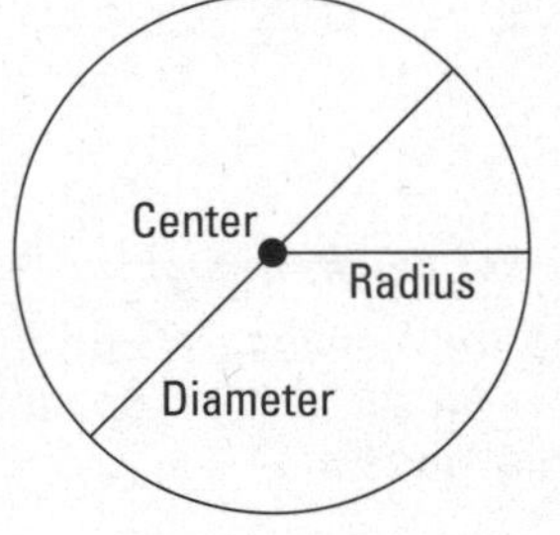

Figure 16-17: Deciphering the parts of a circle.

Illustration by Wiley, Composition Services Graphics

As you can see, the diameter of any circle is made up of one radius plus another radius — that is, two *radii* (pronounced *ray*-dee-eye). This concept gives you the following handy formula:

$$d = 2 \times r$$

For example, given a circle with a radius of 5 millimeters, you can figure out the diameter as follows:

$$d = 2 \times 5\text{mm} = 10\text{mm}$$

Because the circle is an extra-special shape, its perimeter (the length of its "sides") has an extra-special name: the *circumference* (*C* for short). Early mathematicians went to a lot of trouble figuring out how to measure the circumference of a circle. Here's the formula they hit upon:

$$C = 2 \times \pi \times r$$

Note: Because $2 \times r$ is the same as the diameter, you also can write the formula as $C = \pi \times d$.

The symbol π is called *pi* (pronounced "pie"). It's just a number whose approximate value is as follows (the decimal part of pi goes on forever, so you can't get an exact value for pi):

$$\pi \approx 3.14$$

So given a circle with a radius of 5 mm, you can figure out the approximate circumference:

$$C \approx 2 \times 3.14 \times 5\text{mm} = 31.4\text{mm}$$

The formula for the area *(A)* of a circle also uses π:

$$A = \pi \times r^2$$

Here's how to use this formula to find the approximate area of a circle with a radius of 5 mm:

$$A \approx 3.14 \times (5\text{mm})^2 = 3.14 \times 25\text{mm}^2 = 78.5\text{mm}^2$$

Spacing out: Measuring in three dimensions

In three dimensions, the concepts of area has to be tweaked a little. Recall that, in 2-D, the area of a shape is the measurement of what's inside the shape. In 3-D, what's inside a solid is called its *volume.*

The *volume (V)* of a solid is a measurement of the space it occupies, as measured in cubic units, such as cubic inches (in.^{3}), cubic feet (ft.^{3}), cubic meters (m^3), and so forth. (For info on measurement, flip to Chapter 15.)

Finding the volume of solids, however, is something mathematicians love for you to know. In the next sections, I give you the formulas for finding the volumes of a variety of solids.

Cubes

The main measurement of a cube is the length of its side *(s)*. Using this measurement, you can find out the volume of a cube, using the following formula:

$$V = s^3$$

So if the side of a cube is 5 meters, here's how you figure out its volume:

$$V = (5\,\text{m})^3 = 125\,\text{m}^3$$

You can read 125 m^3 as "125 cubic meters" or, less commonly, as "125 meters cubed."

Boxes (rectangular solids)

The three measurements of a box (or rectangular solid) are its length *(l)*, width *(w)*, and height *(h)*. The box pictured in Figure 16-18 has the following measurements: $l = 4$ m, $w = 3$ m, and $h = 2$ m.

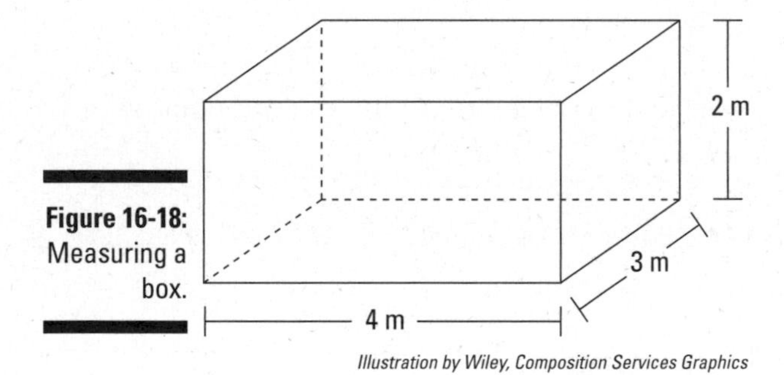

Figure 16-18: Measuring a box.

Illustration by Wiley, Composition Services Graphics

You can find the volume of a box, using the following formula:

$$V = l \times w \times h$$

So here's how to find the volume of the box pictured in this section:

$$V = 4\,\text{m} \times 3\,\text{m} \times 2\,\text{m} = 24\,\text{m}^3$$

Prisms

Finding the volume of a prism (see prisms in Figure 16-7) is easy if you have two measurements. One measurement is the *height* (h) of the prism. The second is the *area of the base* (A_b). The *base* is the polygon that extends vertically from the plane. (In "2-D: Measuring on the flat," earlier, I show you how to find the area of a variety of shapes.)

Here's the formula for finding the volume of a prism:

$$V = A_b \times h$$

For example, suppose a prism has a base with an area of 5 square centimeters and a height of 3 centimeters. Here's how you find its volume:

$$V = 5\,\text{cm}^2 \times 3\,\text{cm} = 15\,\text{cm}^3$$

Notice that the units of measurements (cm^2 and cm) are also multiplied, giving you a result of cm^3.

Cylinders

You find the volume of cylinders the same way you find the area of prisms — by multiplying the area of the base (A_b) by the cylinder's height (h):

$$V = A_b \times h$$

Suppose you want to find the volume of a cylindrical can whose height is 4 inches and whose base is a circle with a radius of 2 inches. First, find the area of the base by using the formula for the area of a circle:

$$\begin{aligned} A &= \pi \times r^2 \\ &\approx 3.14 \times (2\,\text{in.})^2 \\ &= 3.14 \times 4\,\text{in.}^2 \\ &= 12.56\,\text{in.}^2 \end{aligned}$$

This area is approximate because I use 3.14 as an approximate value for π. (***Note:*** In the preceding problem, I use equals signs when a value is equal to whatever comes right before it, and I use "approximately equal to" signs [≈] when I round.)

Now use this area to find the volume of the cylinder:

$$V \approx 12.56\,\text{in.}^2 \times 4\,\text{in.} = 50.24\,\text{in.}^3$$

Notice how multiplying square inches (in.^2) by inches gives a result in cubic inches (in.^3).

Chapter 17

Seeing Is Believing: Graphing as a Visual Tool

In This Chapter

- Making comparisons with a bar graph
- Dividing things up with a pie chart
- Charting change over time with a line graph
- Plotting points and lines on an *xy*-graph

A *graph* is a visual tool for organizing and presenting information about numbers. Most students find graphs relatively easy because they provide a picture to work with rather than just a bunch of numbers. Their simplicity makes graphs show up in newspapers, magazines, business reports, and anywhere clear visual communication is important.

In this chapter, I introduce you to four common styles of graphs: the bar graph, the pie chart, the line graph, and the *xy*-graph. I show you how to read each of these styles of graphs to obtain information. I also show you how to answer the types of questions people may ask when they want to check your understanding.

Looking at Three Important Graph Styles

In this section, I show you how to read and understand three styles of graphs:

- **The bar graph** is best for representing numbers that are independent of each other.
- **The pie chart** allows you to show how a whole is cut up into parts.
- **The line graph** gives you a sense of how numbers change over time.

Bar graph

A bar graph gives you an easy way to compare numbers or values. For example, Figure 17-1 shows a bar graph comparing the performance of five trainers at a fitness center.

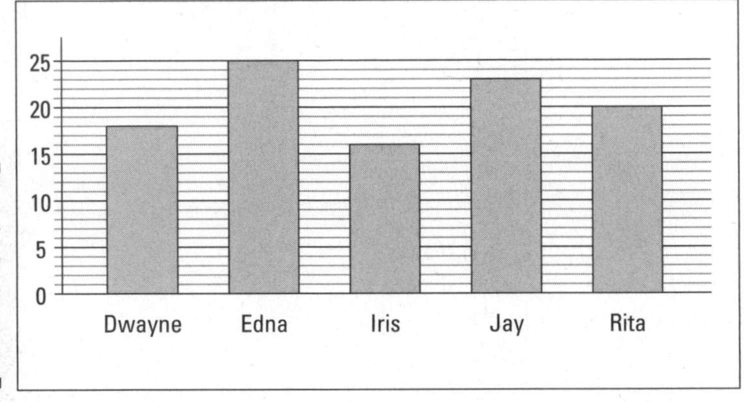

Figure 17-1: The number of new clients recorded this quarter.

Illustration by Wiley, Composition Services Graphics

As you can see from the caption, the graph shows how many new clients each trainer has enrolled this quarter. The advantage of such a graph is that you can see at a glance, for example, that Edna has the most new clients and Iris has the fewest. The bar graph is a good way to represent numbers that are independent of each other. For example, if Iris gets another new client, it doesn't necessarily affect any other trainer's performance.

Reading a bar graph is easy when you get used to it. Here are a few types of questions someone could ask about the bar graph in Figure 17-1:

- **Individual values:** *How many new clients does Jay have?* Find the bar representing Jay's clients and notice that he has 23 new clients.
- **Differences in value:** *How many more clients does Rita have than Dwayne?* Notice that Rita has 20 new clients and Dwayne has 18, so she has 2 more than he does.
- **Totals:** *Together, how many clients do the three women have?* Notice that the three women — Edna, Iris, and Rita — have 25, 16, and 20 new clients, respectively, so they have 61 new clients altogether.

Pie chart

A *pie chart,* which looks like a divided circle, shows you how a whole object is cut up into parts. Pie charts are most often used to represent percentages. For example, Figure 17-2 is a pie chart representing Eileen's monthly expenses.

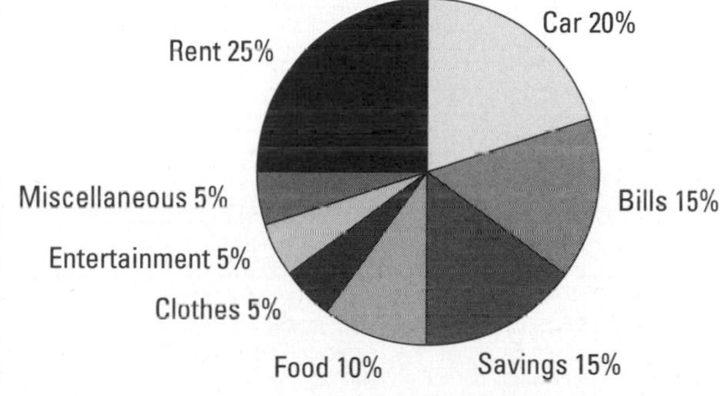

Figure 17-2: Eileen's monthly expenses.

Illustration by Wiley, Composition Services Graphics

You can tell at a glance that Eileen's largest expense is rent and that her second largest is her car. Unlike the bar graph, the pie chart shows numbers that are dependent upon each other. For example, if Eileen's rent increases to 30% of her monthly income, she'll have to decrease her spending in at least one other area.

Here are a few typical questions you may be asked about a pie chart:

- **Individual percentages:** *What percentage of her monthly expenses does Eileen spend on food?* Find the slice that represents what Eileen spends on food and notice that she spends 10% of her income there.
- **Differences in percentages:** *What percentage more does she spend on her car than on entertainment?* Eileen spends 20% on her car but only 5% on entertainment, so the difference between these percentages is 15%.
- **How much a percent represents in terms of dollars:** *If Eileen brings home $2,000 per month, how much does she put away in savings each month?* First notice that Eileen puts 15% every month into savings. So you need to figure out 15% of $2,000. Using your skills from Chapter 12, solve this problem by turning 15% into a decimal and multiplying:

  $$\$2{,}000 \times 0.15 = \$300$$

  So Eileen saves $300 every month.

Line graph

The most common use of a *line graph* is to plot how numbers change over time. For example, Figure 17-3 is a line graph showing last year's sales figures for Tami's Interiors.

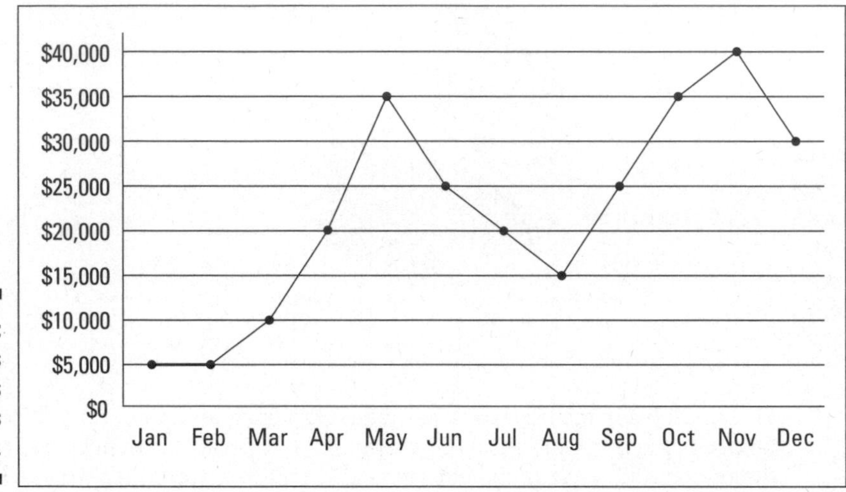

Figure 17-3: Gross receipts for Tami's Interiors.

Illustration by Wiley, Composition Services Graphics

The line graph shows a progression in time. At a glance, you can tell that Tami's business tended to rise strongly at the beginning of the year, drop off during the summer, rise again in the fall, and then drop off again in December.

Here are a few typical questions you may be asked to show that you know how to read a line graph:

- **High or low points and timing:** *In what month did Tami bring in the most revenue, and how much did she bring in?* Notice that the highest point on the graph is in November, when Tami's revenue reached $40,000.
- **Total over a period of time:** *How much did she bring in altogether the last quarter of the year?* A quarter of a year is three months, so the last quarter is the last three months of the year. Tami brought in $35,000 in October, $40,000 in November, and $30,000 in December, so her total receipts for the last quarter add up to $105,000.
- **Greatest change:** *In what month did the business show the greatest gain in revenue as compared with the previous month?* You want to find the line segment on the graph that has the steepest upward slope. This change occurs between April and May, where Tami's revenue increased by $15,000, so her business showed the greatest gain in May.

Using the xy-Graph

When math folks talk about using a graph, they're usually referring to an *xy*-graph (also called the *Cartesian coordinate system*), shown in Figure 17-4. In Chapter 25, I tell you why I believe this graph is one of the ten most important mathematical inventions of all time. You see a lot of this graph when you study algebra, so getting familiar with it now is a good idea.

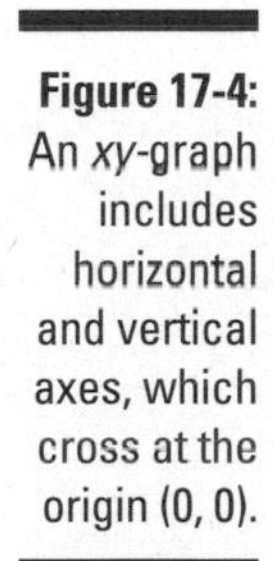

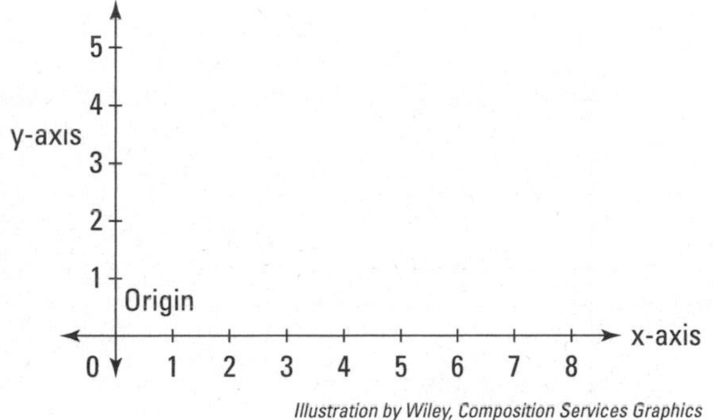

Figure 17-4: An *xy*-graph includes horizontal and vertical axes, which cross at the origin (0, 0).

Illustration by Wiley, Composition Services Graphics

A Cartesian graph is really just two number lines that cross at 0. These number lines are called the *horizontal axis* (also called the *x-axis*) and the *vertical axis* (also called the *y-axis*). The place where these two axes (plural of *axis*) cross is called the *origin*.

Plotting points on an xy-graph

Plotting a point (finding and marking its location) on a graph isn't much harder than finding a point on a number line — after all, a graph is just two number lines put together. (Flip to Chapter 1 for more on using the number line.)

Every point on an *xy*-graph is represented by two numbers in parentheses, separated by a comma, called a set of *coordinates.* To plot any point, start at the origin, where the two axes cross. The first number tells you how far to go to the right (if positive) or left (if negative) along the horizontal axis. The second number tells you how far to go up (if positive) or down (if negative) along the vertical axis.

For example, here are the coordinates of four points called A, B, C, and D:

$A = (2,3)$ $B = (-4,1)$ $C = (0,-5)$ $D = (6,0)$

Figure 17-5 depicts a graph with these four points plotted. Start at the origin, (0, 0). To plot point *A*, count 2 spaces to the right and 3 spaces up. To plot point *B*, count 4 spaces to the left (the negative direction) and then 1 space up. To plot point *C*, count 0 spaces left or right and then count 5 spaces down (the negative direction). And to plot point *D*, count 6 spaces to the right and then 0 spaces up or down.

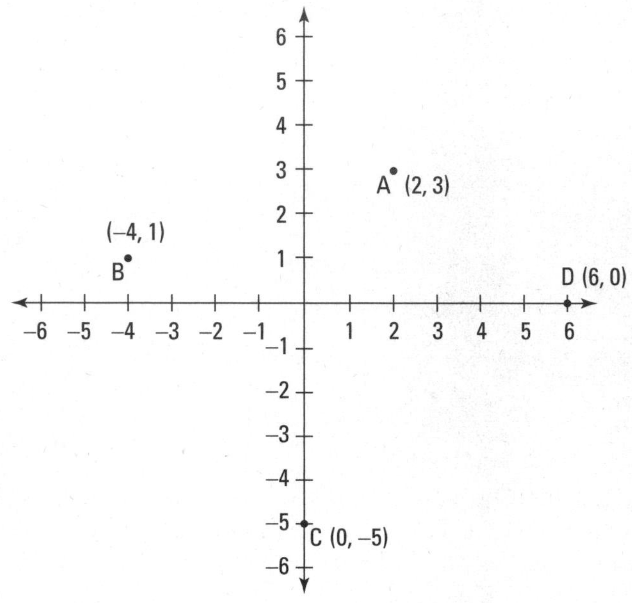

Illustration by Wiley, Composition Services Graphics

Figure 17-5: Points *A*, *B*, *C*, and *D* plotted on an *xy*-graph.

Drawing lines on an xy-graph

When you understand how to plot points on a graph (see the preceding section), you can begin to plot lines and use them to show mathematical relationships.

The examples in this section focus on the number of dollars two people, Xenia and Yanni, are carrying. The horizontal axis represents Xenia's money, and the vertical axis represents Yanni's. For example, suppose you want to draw a line representing this statement:

Xenia has $1 more than Yanni.

Xenia	1	2	3	4	5
Yanni	0	1	2	3	4

Now you have five pairs of points that you can plot on your graph as (Xenia, Yanni): (1,0), (2,1), (3,2), (4,3), and (5,4). Next, draw a straight line through these points, as in Figure 17-6.

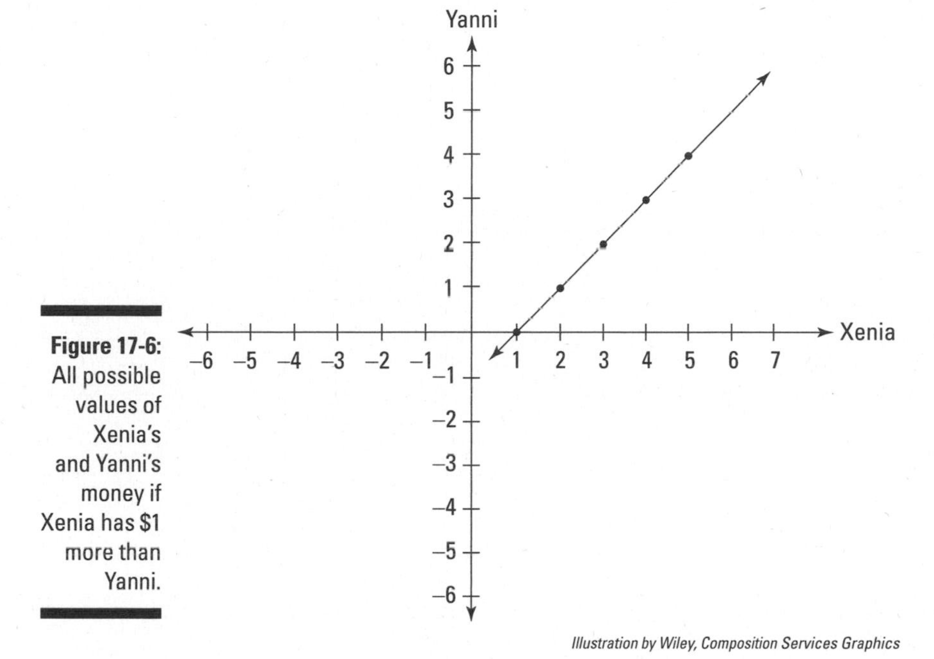

Figure 17-6: All possible values of Xenia's and Yanni's money if Xenia has $1 more than Yanni.

Illustration by Wiley, Composition Services Graphics

This line on the graph represents every possible pair of amounts for Xenia and Yanni. For example, notice how the point (6,5) is on the line. This point represents the possibility that Xenia has $6 and Yanni has $5.

Here's a slightly more complicated example:

> Yanni has $3 more than twice the amount that Xenia has.

Again, start by making the same type of chart as in the preceding example. But this time, if Xenia has $1, then twice that amount is $2, so Yanni has $3 more than that, or $5. Continue in that way to fill in the chart, as follows:

Xenia	1	2	3	4	5
Yanni	5	7	9	11	13

Now plot these five points on the graph and draw a line through them, as in Figure 17-7.

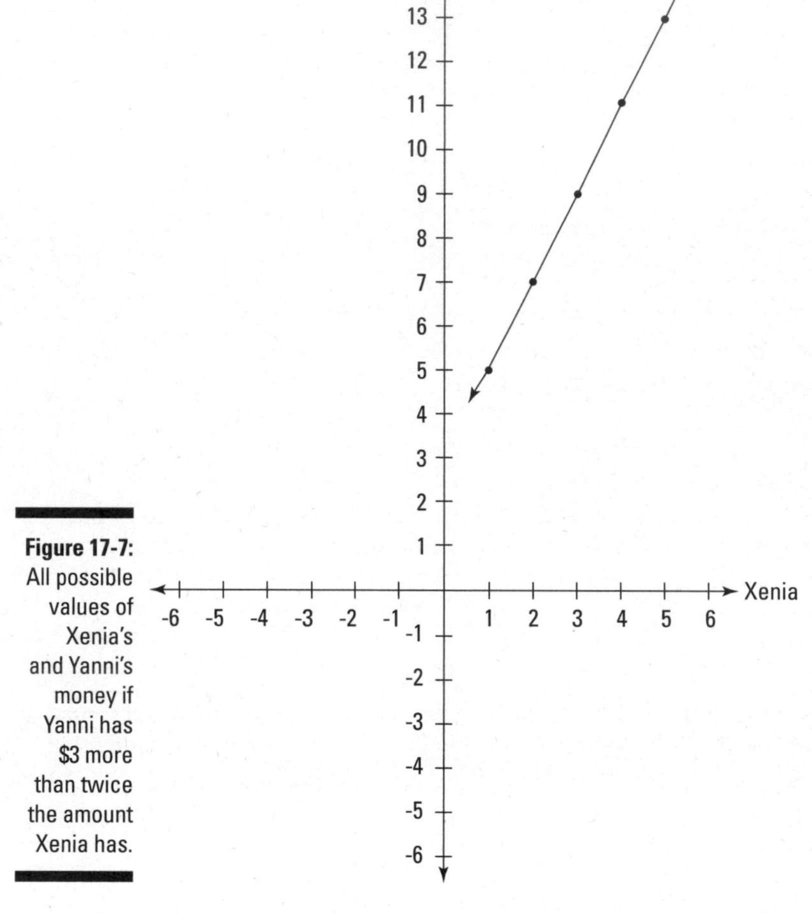

Figure 17-7: All possible values of Xenia's and Yanni's money if Yanni has $3 more than twice the amount Xenia has.

Illustration by Wiley, Composition Services Graphics

As in the other examples, this graph represents all possible values that Xenia and Yanni could have. For example, if Xenia has $7, Yanni has $17.

Chapter 18

Solving Geometry and Measurement Word Problems

In This Chapter

- Solving measurement problems, using conversion chains
- Using a picture to solve geometry problems

In this chapter, I focus on two important types of word problems: measurement problems and geometry problems. In a word problem involving measurement, you're often asked to perform a conversion from one type of unit to another. Sometimes you don't have a conversion equation to solve this type of problem directly, so you need to set up a *conversion chain,* which I discuss in detail in the chapter.

Another common type of word problem requires the geometric formulas that I provide in Chapter 16. Sometimes a geometry word problem gives you a picture to work with. In other cases, you have to draw the picture yourself by reading the problem carefully. Here I give you practice doing both types of problems.

The Chain Gang: Solving Measurement Problems with Conversion Chains

In Chapter 15, I give you a set of basic conversion equations for converting units of measurement. I also show you how to turn these equations into conversion factors — fractions that you can use to convert units. This information is useful as far as it goes, but you may not always have an equation for the exact conversion that you want to perform. For example, how do you convert years to seconds?

For more-complex conversion problems, a good tool is the conversion chain. A *conversion chain* links together a sequence of unit conversions.

Setting up a short chain

Here's a problem that shows you how to set up a short conversion chain to make a conversion you won't find a specific equation for:

> Vendors at the Fragola County Strawberry Festival sold 7 tons of strawberries in a single weekend. How many 1-ounce servings of strawberries is that?

You don't have an equation to convert tons directly to ounces. But you do have one to convert tons to pounds and another to convert pounds to ounces. You can use these equations to build a bridge from one unit to another. So here are the two equations you want to use:

$$1 \text{ ton} = 2{,}000 \text{ lbs.}$$
$$1 \text{ lb.} = 16 \text{ oz.}$$

To convert tons to pounds, note that these fractions equal 1 because the numerator (top number) equals the denominator (bottom number):

$$\frac{1 \text{ ton}}{2000 \text{ lbs.}} \text{ or } \frac{2000 \text{ lbs.}}{1 \text{ ton}}$$

To convert pounds to ounces, note that these fractions equal 1:

$$\frac{1 \text{ lb.}}{16 \text{ oz.}} \text{ or } \frac{16 \text{ oz.}}{1 \text{ lb.}}$$

You could do this conversion in two steps. But when you know the basic idea, you can set up a conversion chain instead to get from tons to ounces:

$$\text{tons} \rightarrow \text{pounds} \rightarrow \text{ounces}$$

So here's how to set up a conversion chain to turn 7 tons into pounds and then into ounces. Because you already have tons on top, you want the tons-and-pounds fraction that puts *ton* on the bottom. And because that fraction puts *pounds* on the top, use the pounds-and-ounces fraction that puts *pound* on the bottom:

$$\frac{7 \text{ tons}}{1} \times \frac{2000 \text{ lbs.}}{1 \text{ ton}} \times \frac{16 \text{ oz.}}{1 \text{ lb.}}$$

The net effect here is to take the expression *7 tons* and multiply it twice by 1, which doesn't change the value of the expression. But now you can cancel out all units of measurement that appear in the numerator of one fraction and the denominator of another:

$$= \frac{7 \cancel{\text{tons}}}{1} \times \frac{2000 \cancel{\text{lbs.}}}{1 \cancel{\text{ton}}} \times \frac{16 \text{ oz.}}{1 \cancel{\text{lb}}.}$$

If any units don't cancel out properly, you probably made a mistake when you set up the chain. Flip the numerator and denominator of one or more of the fractions until the units cancel out the way you want them to.

Now you can simplify the expression:

$$= 7 \times 2{,}000 \times 16 \text{ oz.} = 224{,}000 \text{ oz.}$$

A conversion chain doesn't change the *value* of the expression — just the units of measurement.

Working with more links

When you understand the basic idea of a conversion chain, you can make a chain as long as you like to solve longer problems easily. Here's another example of a problem that uses a time-related conversion chain:

> Jane is exactly 12 years old today. You forgot to get her a present, but you decide that offering her your mathematical skills is the greatest gift of all — you'll recalculate how old she is. Assuming that a year has exactly 365 days, how many seconds old is she?

Here are the conversion equations you have to work with:

$$\begin{aligned} 1 \text{ year} &= 365 \text{ days} \\ 1 \text{ day} &= 24 \text{ hours} \\ 1 \text{ hour} &= 60 \text{ minutes} \\ 1 \text{ minute} &= 60 \text{ seconds} \end{aligned}$$

To solve this problem, you need to build a bridge from years to seconds, as follows:

years → days → hours → minutes → seconds

So set up a long conversion chain, as follows:

$$\frac{12 \text{ years}}{1} \times \frac{365 \text{ days}}{1 \text{ year}} \times \frac{24 \text{ hrs.}}{1 \text{ day}} \times \frac{60 \text{ min.}}{1 \text{ hr.}} \times \frac{60 \text{ sec.}}{1 \text{ min.}}$$

Cancel out all units that appear in both a numerator and a denominator:

$$= \frac{12 \cancel{\text{ years}}}{1} \times \frac{365 \cancel{\text{ days}}}{1 \cancel{\text{ year}}} \times \frac{24 \cancel{\text{ hrs.}}}{1 \cancel{\text{ day}}} \times \frac{60 \cancel{\text{ min.}}}{1 \cancel{\text{ hr.}}} \times \frac{60 \text{ sec.}}{1 \cancel{\text{ min.}}}$$

As you cancel out units, notice that there is a *diagonal* pattern: The numerator (top number) of one fraction cancels with the denominator (bottom number) of the next, and so on.

When the smoke clears, here's what's left:

$$= 12 \times 365 \times 24 \times 60 \times 60 \text{ sec.}$$

This problem requires a bit of multiplication, but the work is no longer confusing:

$$= 378{,}432{,}000 \text{ sec.}$$

The conversion chain from 12 years to 378,432,000 seconds doesn't change the value of the expression — just the unit of measurement.

Pulling equations out of the text

In some word problems, the problem itself gives you a couple of the conversion equations necessary for solving. Take this problem, for example:

> A furlong is $\frac{1}{8}$ of a mile, and a fathom is 2 yards. If I rode my horse 24 furlongs today, how many fathoms did I ride?

This problem gives you two new conversion equations to work with:

- 1 furlong = 1/8 mile
- 1 fathom = 2 yards

It's helpful to remove fractions from the equations before you begin, so here's a more useful version of the first equation:

$$8 \text{ furlongs} = 1 \text{ mile}$$

You also want to remember two other conversions:

$$1 \text{ mile} = 5{,}280 \text{ feet}$$
$$3 \text{ feet} = 1 \text{ yard}$$

Next, build a bridge from furlongs to miles using the conversions available from these equations:

$$\text{furlongs} \rightarrow \text{miles} \rightarrow \text{feet} \rightarrow \text{yards} \rightarrow \text{fathoms}$$

Now you can form your conversion chain. Every unit you want to cancel has to appear once in the numerator and once in the denominator:

$$\frac{24 \text{ furlongs}}{1} \times \frac{1 \text{ mile}}{8 \text{ furlongs}} \times \frac{5280 \text{ feet}}{1 \text{ mile}} \times \frac{1 \text{ yard}}{3 \text{ feet}} \times \frac{1 \text{ fathom}}{2 \text{ yards}}$$

Next, you can cancel out all the units except for fathoms:

$$= \frac{24 \cancel{\text{ furlongs}}}{1} \times \frac{1 \cancel{\text{ mile}}}{8 \cancel{\text{ furlongs}}} \times \frac{5280 \cancel{\text{ feet}}}{1 \cancel{\text{ mile}}} \times \frac{1 \cancel{\text{ yard}}}{3 \cancel{\text{ feet}}} \times \frac{1 \text{ fathom}}{2 \cancel{\text{ yards}}}$$

Another way you can make this problem a little easier is to notice that the number 24 is in the numerator and 3 and 8 are in the denominator. Of course, $3 \times 8 = 24$, so you can cancel out all three of these numbers:

$$= \frac{\cancel{24 \text{ furlongs}}}{1} \times \frac{1 \cancel{\text{ mile}}}{\cancel{8 \text{ furlongs}}} \times \frac{5280 \cancel{\text{ feet}}}{1 \cancel{\text{ mile}}} \times \frac{1 \cancel{\text{ yard}}}{\cancel{3 \text{ feet}}} \times \frac{1 \text{ fathom}}{2 \cancel{\text{ yards}}}$$

At this point, the expression has only two numbers left aside from the 1s, and the fraction's easy to simplify:

$$= \frac{5280}{2} \text{ fathoms} = 2{,}640 \text{ fathoms}$$

As always, the conversion chain from 24 furlongs to 2,640 fathoms doesn't change the value of the expression — just the units of measurement.

Rounding off: Going for the short answer

Sometimes real-life measurements just aren't that accurate. After all, if you measure the length of a football field with your trusty ruler, you're bound to be off an inch or two (or more). When you perform calculations with such measurements, finding the answer to a bunch of decimal places doesn't make sense because the answer's already approximate. Instead, you want to round off your answer to the numbers that are probably correct. Here's a problem that asks you to do just that:

> Heather weighed her new pet hamster, Binky, and found that he weighs 4 ounces. How many grams does Binky weigh, to the nearest whole gram?

This problem requires you to convert from English to metric units, so you need this conversion equation:

$$1\text{ kilogram} \approx 2.20\text{ pounds}$$

Notice that this conversion equation includes only kilograms and pounds, but the problem includes ounces and grams. So to convert from ounces to pounds and from kilograms to grams, here are some equations to help build a bridge between ounces and grams:

$$1\text{ pound} = 16\text{ ounces}$$
$$1\text{ kilogram} = 1{,}000\text{ grams}$$

Your chain will perform the following conversions:

$$\text{ounces} \rightarrow \text{pounds} \rightarrow \text{kilograms} \rightarrow \text{grams}$$

So set up your expression as follows:

$$\frac{4\text{ oz.}}{1} \times \frac{1\text{ lb.}}{16\text{ oz.}} \times \frac{1\text{ kg}}{2.2\text{ lb.}} \times \frac{1000\text{ g}}{1\text{ kg}}$$

As always, after you set up the expression, you can cancel out every unit except for the one you're converting to:

$$= \frac{4\ \cancel{\text{oz.}}}{1} \times \frac{1\ \cancel{\text{lb.}}}{16\ \cancel{\text{oz.}}} \times \frac{1\ \cancel{\text{kg}}}{2.2\ \cancel{\text{lb.}}} \times \frac{1000\text{ g}}{1\ \cancel{\text{kg}}}$$

When you're multiplying a string of fractions, you can make one fraction out of all the numbers. The numbers that were originally in the numerators of fractions remain in the numerator. Similarly, the numbers that were in the denominators remain in the denominator. Then just put a multiplication sign between each pair of numbers.

$$= \frac{4 \times 1000}{16 \times 2.2}\text{ g}$$

At this point, you can begin calculating. But to save some effort, I recommend canceling out common factors. In this case, you cancel out a 4 in the numerator and denominator, changing the 16 in the denominator to a 4:

$$= \frac{\cancel{4} \times 1000}{4\,\cancel{16} \times 2.2}\text{ g}$$

Now you can cancel out another 4 in the numerator and denominator, changing the 1,000 in the numerator to 250:

$$= \frac{\cancel{4} \times \cancel{1000}\,250}{\cancel{4}\,\cancel{16} \times 2.2}\text{ g}$$

At this point, here's what's left:

$$= \frac{250}{2.2}\text{ g}$$

Divide 250 by 2.2 to get your answer:

$$\approx 113.6\text{ g}$$

Notice that I took the division out to one decimal place. Because the number after the decimal point is 6, I need to round up my answer to the next highest gram. (See Chapter 11 for more about rounding decimals.)

So to the nearest gram, Binky weighs 114 grams. As usual, the conversion chain doesn't change the value of the expression — just the unit of measurement.

Solving Geometry Word Problems

Some geometry word problems present you with a picture. In other cases, you have to draw a picture yourself. Sketching figures is always a good idea because it can usually give you an idea of how to proceed. The following sections present you with both types of problems. (To solve these word problems, you need some of the geometry formulas I discuss in Chapter 16.)

Working from words and images

Sometimes you have to interpret a picture to solve a word problem. Read the problem carefully, recognize shapes in the drawing, pay attention to labels, and use whatever formulas you have to help you answer the question. In this problem, you get to work with a picture.

> Mr. Dennis is a farmer with two teenage sons. He gave them a rectangular piece of land with a creek running through it diagonally, as shown in Figure 18-1. The elder boy took the larger area, and the younger boy took the smaller. What is the area of each boy's land in square feet?

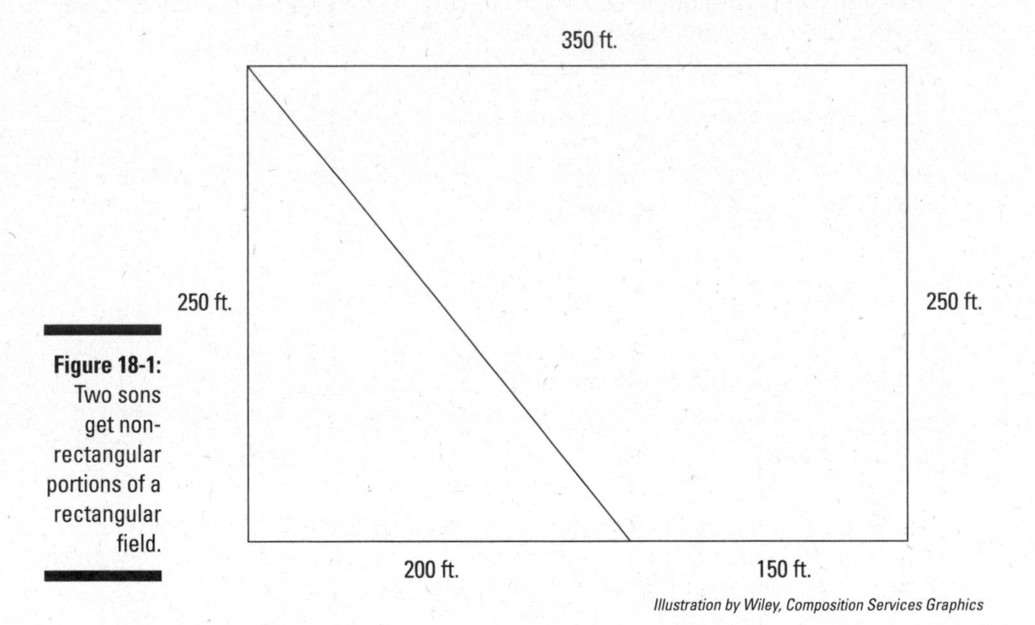

Figure 18-1: Two sons get non-rectangular portions of a rectangular field.

Illustration by Wiley, Composition Services Graphics

To find the area of the smaller triangular plot, use the formula for the area of a triangle, where A is the area, b is the base, and h is the height:

$$A = \frac{1}{2}(b \times h)$$

The whole piece of land is a rectangle, so you know that the corner the triangle shares with the rectangle is a right angle. Therefore, you know that the sides labeled 200 feet and 250 feet are the base and height. Find the area of this plot by plugging the base and height into the formula:

$$A = \frac{200 \text{ feet} \times 250 \text{ feet}}{2}$$

To make this calculation a little easier, notice that you can cancel a factor of 2 from the numerator and denominator:

$$A = \frac{^{100}\cancel{200} \text{ feet} \times 250 \text{ feet}}{\cancel{2}} = 25{,}000 \text{ square feet}$$

The shape of the remaining area is a trapezoid. You can find its area by using the formula for a trapezoid, but there's an easier way. Because you know the area of the triangular plot, you can use this word equation to find the area of the trapezoid:

area of trapezoid = area of whole plot – area of triangle

To find the area of the whole plot, remember the formula for the area of a rectangle. Plug its length and width into the formula:

$$A = \text{length} \times \text{width}$$
$$A = 350 \text{ ft.} \times 250 \text{ ft.}$$
$$A = 87{,}500 \text{ square ft.}^2$$

Now just substitute the numbers that you know into the word equation you set up:

$$\text{area of trapezoid} = 87{,}500 \text{ square feet} - 25{,}000$$
$$= 62{,}500 \text{ square feet}$$

So the area of the elder boy's land is 62,500 square feet, and the area of the younger boy's land is 25,000 square feet.

Breaking out those sketching skills

Geometry word problems may not make much sense until you draw some pictures. Here's an example of a geometry problem without a picture provided:

In Elmwood Park, the flagpole is due south of the swing set and exactly 20 meters due west of the tree house. If the area of the triangle made by the flagpole, the swing set, and the tree house is 150 square meters, what is the distance from the swing set to the tree house?

This problem is bound to be confusing until you draw a picture of what it's telling you. Start with the first sentence, depicted in Figure 18-2. As you can see, I've drawn a right triangle whose corners are the swing set *(S)*, the flagpole *(F)*, and the tree house *(T)*. I've also labeled the distance from the flagpole to the tree house as 20 meters.

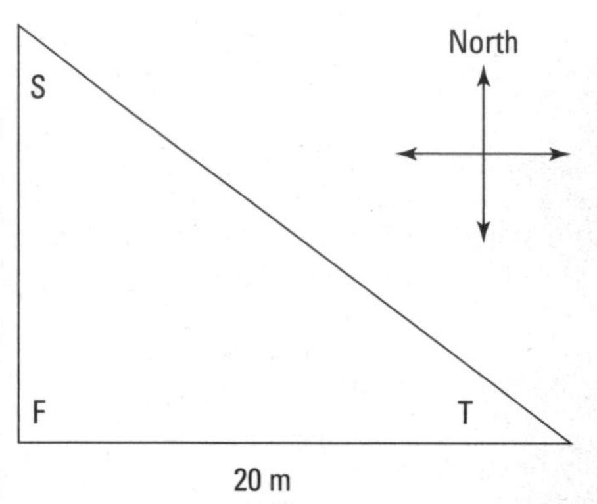

Figure 18-2: A labeled sketch shows the important information in a word problem.

Illustration by Wiley, Composition Services Graphics

The next sentence tells you the area of this triangle:

$$A = 150\,\text{m}^2$$

Now you're out of information, so you need to remember anything you can from geometry. Because you know the area of the triangle, you may find the formula for the area of a triangle helpful:

$$A = \frac{1}{2}(b \times h)$$

Here b is the base and h is the height. In this case, you have a right triangle, so the base is the distance from F to T, and the height is the distance from S to F. So you already know the area of the triangle, and you also know the length of the base. Fill in the equation:

$$150 = \frac{1}{2}(20 \times h)$$

You can now solve this equation for h. Start by simplifying:

$$150 = 10 \times h$$
$$15 = h$$

Now you know that the height of the triangle is 15 meters, so you can add this information to your picture (see Figure 18-3).

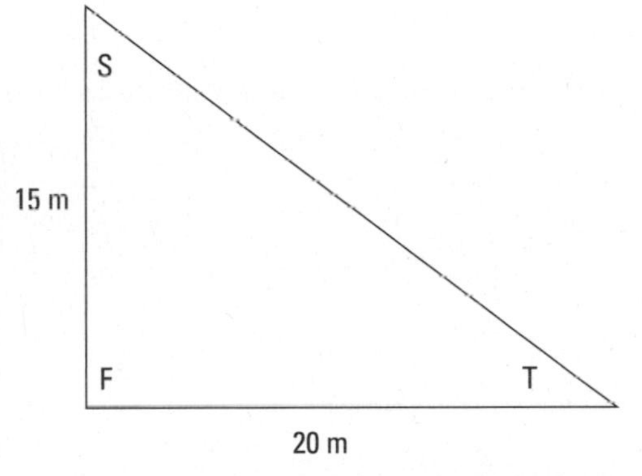

Figure 18-3: Update the labels in your sketch as you work through the problem.

Illustration by Wiley, Composition Services Graphics

To solve the problem, though, you still need to find out the distance from S to T. Because this is a right triangle, you can use the Pythagorean theorem to figure out the distance:

$$a^2 + b^2 = c^2$$

Remember that a and b are the lengths of the short sides, and c is the length of the longest side, called the *hypotenuse*. (See Chapter 16 for more on the Pythagorean theorem.) You can substitute numbers into this formula and solve, as follows:

$$15^2 + 20^2 = c^2$$
$$225 + 400 = c^2$$
$$625 = c^2$$
$$\sqrt{625} = \sqrt{c^2}$$
$$25 = c$$

So the distance from the swing set to the tree house is 25 meters.

Chapter 19

Figuring Your Chances: Statistics and Probability

In This Chapter

- Knowing how statistics works with both qualitative and quantitative data
- Finding out how to calculate a percentage and the mode of a sample
- Calculating the mean and median
- Finding the probability of an event

Statistics and probability are two of the most important and widely used applications of math. They're applicable to virtually every aspect of the real world — business, biology, city planning, politics, meteorology, and many more areas of study. Even physics, once thought to be devoid of uncertainty, now relies on probability.

In this chapter, I give you a basic understanding of these two mathematical ideas. First, I introduce you to statistics and the important distinction between qualitative and quantitative data. I show you how to work with both types of data to find meaningful answers. Then I give you the basics of probability. I show you how the probability that an event will occur is always a number from 0 to 1 — that is, usually a fraction, decimal, or percent. After that, I demonstrate how to build this number by counting both favorable outcomes and possible outcomes. Finally, I put these ideas to work by showing you how to calculate the probability of tossing coins.

Gathering Data Mathematically: Basic Statistics

Statistics is the science of gathering and drawing conclusions from data, which is information that's measured objectively in an unbiased, reproducible way.

An individual *statistic* is a conclusion drawn from this data. Here are some examples:

- The average working person drinks 3.7 cups of coffee every day.
- Only 52% of students who enter law school actually graduate.
- The cat is the most popular pet in the United States.
- In the last year, the cost of a high-definition TV dropped by an average of $575.

Statisticians do their work by identifying a population that they want to study: working people, law students, pet owners, buyers of electronics, whoever. Because most populations are far too large to work with, a statistician collects data from a smaller, randomly selected sample of this population. Much of statistics concerns itself with gathering data that's reliable and accurate. You can read all about this idea in *Statistics For Dummies,* 2nd Edition, by Deborah J. Rumsey (Wiley).

In this section, I give you a short introduction to the more mathematical aspects of statistics.

Understanding differences between qualitative and quantitative data

Data — the information used in statistics — can be either qualitative or quantitative. *Qualitative data* divides a data set (the pool of data that you've gathered) into discrete chunks based on a specific attribute. For example, in a class of students, qualitative data can include

- Each child's gender
- His or her favorite color
- Whether he or she owns at least one pet
- How he or she gets to and from school

You can identify qualitative data by noticing that it links an attribute — that is, a quality — to each member of the data set. For example, four attributes of Emma are that she's female, her favorite color is green, she owns a dog, and she walks to school.

On the other hand, *quantitative data* provides numerical information — that is, information about quantities, or amounts. For example, quantitative data on this same classroom of students can include the following:

- Each child's height in inches
- Each child's weight in pounds

- The number of siblings each child has
- The number of words each child spelled correctly on the most recent spelling test

You can identify quantitative data by noticing that it links a number to each member of the data set. For example, Carlos is 55 inches tall, weighs 68 pounds, has three siblings, and spelled 18 words correctly.

Working with qualitative data

Qualitative data usually divides a sample into discrete chunks. As my sample — which is purely fictional — I use 25 children in Sister Elena's fifth-grade class. For example, suppose all 25 children in Sister Elena's class answer the three yes/no questions in Table 19-1.

Table 19-1 Sister Elena's Fifth-Grade Survey

Question	*Yes*	*No*
Are you an only child?	5	20
Do you own any pets?	14	11
Do you take the bus to school?	16	9

The students also answer the question "What is your favorite color?" with the results in Table 19-2.

Table 19-2 Favorite Colors in Sister Elena's Class

Color	*Number of Students*	*Color*	*Number of Students*
Blue	8	Orange	1
Red	6	Yellow	1
Green	5	Gold	1
Purple	3		

Even though the information that each child provided is non-numerical, you can handle it numerically by counting how many students made each response and working with these numbers.

Given this information, you can now make informed statements about the students in this class just by reading the charts. For instance,

- Exactly 20 children have at least one brother or sister.
- Nine children don't take the bus to school.
- Only one child's favorite color is yellow.

Playing the percentages

You can make more sophisticated statistical statements about qualitative data by finding out the percentage of the sample that has a specific attribute. Here's how you do so:

1. **Write a statement that includes the number of members who share that attribute and the total number in the sample.**

   Suppose you want to know what percentage of students in Sister Elena's class are only children. The chart tells you that 5 students have no siblings, and you know that 25 kids are in the class. So you can begin to answer this question as follows:

   Five out of 25 children are only children.

2. **Rewrite this statement, turning the numbers into a fraction:**

   $$\frac{\text{number who share attribute}}{\text{number in sample}} = \frac{5}{25}$$

   In the example, $\frac{5}{25}$ of the children are only children.

3. **Turn the fraction into a percent, using the method I show you in Chapter 12.**

   You find that $\frac{5}{25} = \frac{1}{5} = 0.2$, so 20% of the children are only children.

Similarly, suppose you want to find out what percentage of children take the bus to school. This time, the chart tells you that 16 children take the bus, so you can write this statement:

Sixteen out of 25 children take the bus to school.

Now rewrite the statement as follows:

$\frac{16}{25}$ of the children take the bus to school.

Finally, turn this fraction into a percent: $16 \div 25 = 0.64$, which equals 64%, so

64% of the children take the bus to school.

Getting into the mode

The *mode* tells you the most popular answer to a statistical question. For example, in the poll of Sister Elena's class (see Tables 19-1 and 19-2), the mode groups are children who

- Have at least one brother or sister (20 students)
- Own at least one pet (14 students)
- Take the bus to school (16 students)
- Chose blue as their favorite color (8 students)

When a question divides a data set into two parts (as with all yes/no questions), the mode group represents more than half of the data set. But when a question divides a data set into more than two parts, the mode doesn't necessarily represent more than half of the data set.

For example, 14 children own at least one pet, and the other 11 children don't own one. So the mode group — children who own a pet — is more than half the class. But 8 of the 25 children chose blue as their favorite color. So even though this is the mode group, fewer than half the class chose this color.

With a small sample, you may have more than one mode — for example, perhaps the number of students who like red is equal to the number who like blue. However, getting multiple modes isn't usually an issue with a larger sample because it becomes less likely that exactly the same number of people will have the same preference.

Working with quantitative data

Quantitative data assigns a numerical value to each member of the sample. As my sample — again, fictional — I use five members of Sister Elena's basketball team. Suppose that the information in Table 19-3 has been gathered about each team member's height and most recent spelling test.

Table 19-3 Height and Spelling Test Scores

Student	*Height in Inches*	*Number of Words Spelled Correctly*
Carlos	55	18
Dwight	60	20
Patrick	59	14
Tyler	58	17
William	63	18

In this section, I show you how to use this information to find the mean and median for both sets of data. Both terms refer to ways to calculate the average value in a quantitative data set. An *average* gives you a general idea of where most individuals in a data set fall so you know what kinds of results are standard. For example, the average height of Sister Elena's fifth-grade class is probably less than the average height of the Los Angeles Lakers. As I show you in the sections that follow, an average can be misleading in some cases, so knowing when to use the mean versus the median is important.

Finding the mean

REMEMBER

The mean is the most commonly used average. In fact, when most people use the word *average,* they're referring to the mean. Here's how you find the mean of a set of data:

1. **Add up all the numbers in that set.**

   For example, to find the average height of the five team members, first add up all their heights:

   $$55 + 60 + 59 + 58 + 63 = 295$$

2. **Divide this result by the total number of members in that set.**

   Divide 295 by 5 (that is, by the total number of boys on the team):

   $$295 \div 5 = 59$$

   So the mean height of the boys on Sister Elena's team is 59 inches.

This procedure is summed up (so to speak) in simple formula:

$$\text{mean} = \frac{\text{sum of values}}{\text{number of values}}$$

You can use this formula to find the mean number of words that the boys spelled correctly. To do this, plug the number of words that each boy spelled correctly into the top part of the formula, and then plug the number of boys in the group into the bottom part:

$$\text{mean} = \frac{18 + 20 + 14 + 17 + 18}{5}$$

Now simplify to find the result:

$$= \frac{87}{5} = 17.4$$

As you can see, when you divide, you end up with a decimal in your answer. If you round to the nearest whole word, the mean number of words that the five boys spelled correctly is about 17 words. (For more information about rounding, see Chapter 2.)

The mean can be misleading when you have a strong skew in data — that is, when the data has many low values and a few very high ones, or vice versa.

For example, suppose that the president of a company tells you, "The average salary in my company is $200,000 a year!" But on your first day at work, you find out that the president's salary is $19,010,000 and each of his 99 employees earns $10,000. To find the mean, first plug the total salaries ($19,010,000 for the president plus $10,000 for each of 99 employees) into the top of the formula. Next, plug the number of employees (100) into the bottom:

$$\text{mean} = \frac{\$19,010,000 + (\$10,000 \times 99)}{100}$$

Now calculate:

$$= \frac{\$19,010,000 + \$990,000}{100} = \frac{\$20,000,000}{100} = \$200,000$$

So the president didn't lie. However, the skew in salaries resulted in a misleading mean.

Finding the median

When data values are skewed (when a few very high or very low numbers differ significantly from the rest of the data), the median can give you a more accurate picture of what's standard. Here's how to find the median of a set of data:

1. **Arrange the set from lowest to highest.**

   To find the median height of the boys in Table 19-3, arrange their five heights in order from lowest to highest.

   55 58 <u>59</u> 60 63

2. **Choose the middle number.**

   The middle value, 59 inches, is the median average height.

To find the median number of words that the boys spelled correctly (refer to Table 19-3), arrange their scores in order from lowest to highest:

14 17 <u>18</u> 18 20

This time, the middle value is 18, so 18 is the median score.

If you have an even number of values in the data set, put the numbers in order and find the mean of the *two middle numbers* in the list (see the preceding section for details on the mean). For instance, consider the following:

2   3   5   7   9   11

The two center numbers are 5 and 7. Add them together to get 12, and then divide by 2 to get their mean. The median in this list is 6.

Now recall the company president who makes $19,010,000 a year and his 99 employees who each earn $10,000. Here's how this data looks:

10,000   10,000   10,000 ...   10,000   19,010,000

As you can see, if you wrote out all 100 salaries, the center numbers would obviously both be 10,000. The median salary is $10,000, and this result is much more reflective of what you'd probably earn if you worked at this company.

Looking at Likelihoods: Basic Probability

Probability is the mathematics of deciding how likely an event is to occur. For example,

- What's the likelihood that the lottery ticket I bought will win?
- What's the likelihood that my new car will need repairs before the warranty runs out?
- What's the likelihood that more than 100 inches of snow will fall in Manchester, New Hampshire, this winter?

Probability has a wide variety of applications in insurance, weather prediction, biological sciences, and even physics.

The study of probability started hundreds of years ago when a group of French noblemen began to suspect that math could help them turn a profit, or at least not lose so heavily, in the gambling halls they frequented.

You can read all about the details of probability in *Probability For Dummies,* by Deborah J. Rumsey (Wiley). In this section, I give you a little taste of this fascinating subject.

The Silver standard

Probability can be a powerful tool for predicting weather patterns, sports events, and election results. In his bestselling book *The Signal and the Noise*, Nate Silver discusses how statistical modeling, when done correctly, can permit mathematicians to peer into the future with spooky accuracy. He also discusses why a lot of apparently scientific predictions go wrong. Silver's work is cutting edge, and he does a good job of explaining what statisticians do without too much jargon or complicated equations. Check him out!

Figuring the probability

The *probability* that an event will occur is a fraction whose numerator (top number) and denominator (bottom number) are as follows (for more on fractions, flip to Chapter 9):

$$\text{probability} = \frac{\text{target outcomes}}{\text{total outcomes}}$$

In this case, the number of *target outcomes* (or *successes*) is simply the number of outcomes in which the event you're examining does happen. In contrast, the number of *total outcomes* (or *sample space*) is the number of outcomes that *can* happen.

For example, suppose you want to know the probability that a tossed coin will land heads up. Notice that there are two total outcomes (heads or tails), but only one of these outcomes is the target — the outcome in which heads comes up. To find the probability of this event, make a fraction as follows:

$$\text{probability} = \frac{1}{2}$$

So the probability that the coin will land heads up is $\frac{1}{2}$.

So what's the probability that, when you roll a die, the number 3 will land face up? To figure this one out, notice that there are *six* total outcomes (1, 2, 3, 4, 5, and 6), but in only *one* of these does 3 land face up. To find the probability of this outcome, make a fraction as follows:

$$\text{probability} = \frac{1}{6}$$

So the probability that the number 3 will land face up is $\frac{1}{6}$.

And what's the probability that, if you pick a card at random from a deck, it'll be an ace? To figure this out, notice that there are *52* total outcomes (one for each card in the deck), but in only *4* of these do you pick an ace. So

$$\text{probability} = \frac{4}{52}$$

So the probability that you'll pick an ace is $\frac{4}{52}$, which reduces to $\frac{1}{13}$ (see Chapter 9 for more on reducing fractions).

Probability is always a number from 0 to 1. When the probability of an outcome is 0, the outcome is *impossible.* When the probability of an outcome is 1, the outcome is *certain.*

Oh, the possibilities! Counting outcomes with multiple coins

Although the basic probability formula isn't difficult, sometimes finding the numbers to plug into it can be tricky. One source of confusion is in counting the number of outcomes, both target and total. In this section, I focus on tossing coins.

When you flip a coin, you can generally get two total outcomes: heads or tails. When you flip two coins at the same time — say, a penny and a nickel — you can get four total outcomes:

Outcome	***Penny***	***Nickel***
#1	Heads	Heads
#2	Heads	Tails
#3	Tails	Heads
#4	Tails	Tails

When you flip three coins at the same time — say, a penny, a nickel, and a dime — eight outcomes are possible:

Outcome	***Penny***	***Nickel***	***Dime***
#1	Heads	Heads	Heads
#2	Heads	Heads	Tails
#3	Heads	Tails	Heads
#4	Heads	Tails	Tails

Outcome	*Penny*	*Nickel*	*Dime*
#5	Tails	Heads	Heads
#6	Tails	Heads	Tails
#7	Tails	Tails	Heads
#8	Tails	Tails	Tails

Notice the pattern: Every time you add a coin, the number of total outcomes doubles. So if you flip six coins, here's how many total outcomes you have:

$$2 \times 2 \times 2 \times 2 \times 2 \times 2 = 64$$

The number of total outcomes equals the number of outcomes per coin (2) raised to the number of coins (6): Mathematically, you have $2^6 = 64$.

Here's a handy formula for calculating the number of outcomes when you're flipping, shaking, or rolling multiple coins, dice, or other objects at the same time:

$$\text{total outcomes} = \text{number of outcomes per object}^{\text{number of objects}}$$

Suppose you want to find the probability that six tossed coins will all fall heads up. To do this, you want to build a fraction, and you already know that the denominator — the number of total outcomes — is 64. Only one outcome is the target outcome, so the numerator is 1:

$$\text{probability} = \frac{1}{64}$$

So the probability that six tossed coins will all fall heads up is $\frac{1}{64}$.

Here's a more subtle question: What's the probability that exactly five out of six tossed coins will all fall heads up? Again, you're building a fraction, and you already know that the denominator is 64. To find the numerator (target outcomes), think about it this way: If the first coin falls tails up, then all the rest must fall heads up. If the second coin falls tails up, then again all the rest must fall heads up. This is true of all six coins, so you have six target outcomes:

$$\text{probability} = \frac{6}{64}$$

Therefore, the probability that exactly five out of six coins will fall heads up is $\frac{6}{64}$, which reduces to $\frac{3}{32}$ (see Chapter 9 for more on reducing fractions).

Chapter 20

Setting Things Up with Basic Set Theory

In This Chapter

- Defining a set and its elements
- Understanding subsets and the empty set
- Knowing the basic operations on sets, including union and intersection

A *set* is just a collection of things. But in their simplicity, sets are profound. At the deepest level, set theory is the foundation for everything in math.

Set theory provides a way to talk about collections of numbers, such as even numbers, prime numbers, or counting numbers, with ease and clarity. It also gives rules for performing calculations on sets that become useful in higher math. For these reasons, set theory becomes more important the higher up you go the math food chain — especially when you begin writing mathematical proofs. Studying sets can also be a nice break from the usual math stuff you work with.

In this chapter, I show you the basics of set theory. First, I show you how to define sets and their elements and how you can tell when two sets are equal. I also show you the simple idea of a set's cardinality. Next, I discuss subsets and the all-important empty set (∅). After that, I discuss four operations on sets: union, intersection, relative complement, and complement.

Understanding Sets

A *set* is a collection of things, in any order. These things can be buildings, earmuffs, lightning bugs, numbers, qualities of historical figures, names you call your little brother, whatever.

You can define a set in a few main ways:

- **Placing a list of the elements of the set in braces { }:** You can simply list everything that belongs in the set. When the set is too large, you use an ellipsis (...) to indicate elements of the set not mentioned. For example, to list the set of numbers from 1 to 100, you can write {1, 2, 3, ..., 100}. To list the set of all the counting numbers, you can write {1, 2, 3, ...}.
- **Using a verbal description:** If you use a verbal description of what the set includes, make sure the description is clear and unambiguous so you know exactly what's in the set and what isn't. For instance, the set of the four seasons is pretty clear-cut, but you may run into some debate on the set of words that describe my cooking skills because different people have different opinions.
- **Writing a mathematical rule (set-builder notation):** In later algebra, you can write an equation that tells people how to calculate the numbers that are part of a set. Check out *Algebra II For Dummies,* by Mary Jane Sterling (Wiley), for details.

Sets are usually identified with capital letters to keep them distinct from variables in algebra, which are usually small letters. (Chapter 21 talks about using variables.)

The best way to understand sets is to begin working with them. For example, here I define three sets:

A = {Empire State Building, Eiffel Tower, Roman Colosseum}

B = {Albert Einstein's intelligence, Marilyn Monroe's talent, Joe DiMaggio's athletic ability, Sen. Joseph McCarthy's ruthlessness}

C = the four seasons of the year

Set A contains three tangible objects: famous works of architecture. Set B contains four intangible objects: attributes of famous people. And set C also contains intangible objects: the four seasons. Set theory allows you to work with either tangible or intangible objects, provided that you define your set properly. In the following sections, I show you the basics of set theory.

Elementary, my dear: Considering what's inside sets

The things contained in a set are called *elements* (also known as *members*). Consider the first two sets I define in the section intro:

A = {Empire State Building, Eiffel Tower, Roman Colosseum}

B = {Albert Einstein's intelligence, Marilyn Monroe's talent, Joe DiMaggio's athletic ability, Sen. Joseph McCarthy's ruthlessness}

The Eiffel Tower is an element of A, and Marilyn Monroe's talent is an element of B. You can write these statements using the symbol ∈, which means "is an element of":

Eiffel Tower ∈ A

Marilyn Monroe's talent ∈ B

However, the Eiffel Tower is not an element of B. You can write this statement using the symbol ∉, which means "is not an element of":

Eiffel Tower ∉ B

These two symbols become more common as you move higher in your study of math. The following sections discuss what's inside those braces and how some sets relate to each other.

Cardinality of sets

The *cardinality* of a set is just a fancy word for the number of elements in that set.

When A is {Empire State Building, Eiffel Tower, Roman Colosseum}, it has three elements, so the cardinality of A is three. Set B, which is {Albert Einstein's intelligence, Marilyn Monroe's talent, Joe DiMaggio's athletic ability, Sen. Joseph McCarthy's ruthlessness}, has four elements, so the cardinality of B is four.

Equal sets

If two sets list or describe the exact same elements, the sets are equal (you can also say they're *identical* or *equivalent*). The order of elements in the sets doesn't matter. Similarly, an element may appear twice in one set, but only the distinct elements need to match.

Suppose I define some sets as follows:

C = the four seasons of the year

D = {spring, summer, fall, winter}

E = {fall, spring, summer, winter}

F = {summer, summer, summer, spring, fall, winter, winter, summer}

Set C gives a clear rule describing a set. Set D explicitly lists the four elements in C. Set E lists the four seasons in a different order. And set F lists the four seasons with some repetition. Thus, all four sets are equal. As with numbers, you can use the equals sign to show that sets are equal:

C = D = E = F

Subsets

When all the elements of one set are completely contained in a second set, the first set is a subset of the second. For example, consider these sets:

C = {spring, summer, fall, winter}

G = {spring, summer, fall}

As you can see, every element of G is also an element of C, so G is a subset of C. The symbol for subset is ⊂, so you can write the following:

G ⊂ C

Every set is a subset of itself. This idea may seem odd until you realize that all the elements of any set are obviously contained in that set.

Empty sets

The *empty set* — also called the *null set* — is a set that has no elements:

H = {}

As you can see, I define H by listing its elements, but I haven't listed any, so H is empty. The symbol ∅ is used to represent the empty set. So H = ∅.

You can also define an empty set by using a rule. For example,

I = types of roosters that lay eggs

Clearly, roosters are male and, therefore, can't lay eggs, so this set is empty.

You can think of ∅ as nothing. And because nothing is always nothing, there's only one empty set. All empty sets are equal to each other, so in this case, H = I.

Furthermore, ∅ is a subset of every other set (the preceding section discusses subsets), so the following statements are true:

∅ ⊂ A

∅ ⊂ B

∅ ⊂ C

This concept makes sense when you think about it. Remember that ∅ has no elements, so technically, every element in ∅ is in every other set.

Sets of numbers

One important use of sets is to define sets of numbers. As with all other sets, you can do so either by listing the elements or by verbally describing a rule that clearly tells you what's included in the set and what isn't. For example, consider the following sets:

J = {1, 2, 3, 4, 5}

K = {2, 4, 6, 8, 10,...}

L = the set of counting numbers

My definitions of J and K list their elements explicitly. Because K is infinitely large, you need to use an ellipsis (...) to show that this set goes on forever. The definition of L is a description of the set in words.

I discuss some especially significant sets of numbers in Chapter 25.

Performing Operations on Sets

In arithmetic, the Big Four operations (adding, subtracting, multiplying, and dividing) allow you to combine numbers in various ways (see Chapters 3 and 4 for more information). Set theory also has four important operations: union, intersection, relative complement, and complement. You'll see more of these operations as you move on in your study of math.

Here are definitions for three sets of numbers:

P = {1, 7}

Q = {4, 5, 6}

R = {2, 4, 6, 8, 10}

In this section, I use these three sets and a few others to discuss the four set operations and show you how they work. (***Note:*** Within equations, I relist the elements, replacing the names of the sets with their equivalent in braces. Therefore, you don't have to flip back and forth to look up what each set contains.)

Union: Combined elements

The union of two sets is the set of their *combined* elements. For example, the union of {1, 2} and {3, 4} is {1, 2, 3, 4}. The symbol for this operation is ∪, so

$$\{1, 2\} \cup \{3, 4\} = \{1, 2, 3, 4\}$$

Similarly, here's how to find the union of P and Q:

$$P \cup Q = \{1, 7\} \cup \{4, 5, 6\} = \{1, 4, 5, 6, 7\}$$

When two sets have one or more elements in common, these elements appear only once in their union set. For example, consider the union of Q and R. In this case, the elements 4 and 6 are in both sets, but each of these numbers appears once in their union:

$$Q \cup R = \{4, 5, 6\} \cup \{2, 4, 6, 8, 10\} = \{2, 4, 5, 6, 8, 10\}$$

The union of any set with itself is itself:

$$P \cup P = P$$

Similarly, the union of any set with ∅ (see the earlier "Empty sets" section) is itself:

$$P \cup \varnothing = P$$

Intersection: Elements in common

The intersection of two sets is the set of their common elements (the elements that appear in both sets). For example, the intersection of {1, 2, 3} and {2, 3, 4} is {2, 3}. The symbol for this operation is ∩. You can write the following:

$$\{1, 2, 3\} \cap \{2, 3, 4\} = \{2, 3\}$$

Similarly, here's how to write the intersection of Q and R:

$$Q \cap R = \{4, 5, 6\} \cap \{2, 4, 6, 8, 10\} = \{4, 6\}$$

When two sets have no elements in common, their intersection is the empty set (∅):

$$P \cap Q = \{1, 7\} \cap \{4, 5, 6\} = \varnothing$$

The intersection of any set with itself is itself:

$$P \cap P = P$$

But the intersection of any set with ∅ is ∅:

$$P \cap \varnothing = \varnothing$$

Relative complement: Subtraction (sorta)

The relative complement of two sets is an operation similar to subtraction. The symbol for this operation is the minus sign (–). Starting with the first set, you remove every element that appears in the second set to arrive at their relative complement. For example,

$$\{1, 2, 3, 4, 5\} - \{1, 2, 5\} = \{3, 4\}$$

Similarly, here's how to find the relative complement of R and Q. Both sets share a 4 and a 6, so you have to remove those elements from R:

$$R - Q = \{2, 4, 6, 8, 10\} - \{4, 5, 6\} = \{2, 8, 10\}$$

Note that the reversal of this operation gives you a different result. This time, you remove the shared 4 and 6 from Q:

$$Q - R = \{4, 5, 6\} - \{2, 4, 6, 8, 10\} = \{5\}$$

Like subtraction in arithmetic, the relative complement is not a commutative operation. In other words, order is important. (See Chapter 4 for more on commutative and non-commutative operations.)

Complement: Feeling left out

The complement of a set is everything that isn't in that set. Because "everything" is a difficult concept to work with, you first have to define what you mean by "everything" as the universal set (U). For example, suppose you define the universal set like this:

$$U = \{0, 1, 2, 3, 4, 5, 6, 7, 8, 9\}$$

Now, here are a couple of sets to work with:

$$M = \{1, 3, 5, 7, 9\}$$
$$N = \{6\}$$

The complement of each set is the set of every element in U that isn't in the original set:

$$U - M = \{0, 1, 2, 3, 4, 5, 6, 7, 8, 9\} - \{1, 3, 5, 7, 9\} = \{0, 2, 4, 6, 8\}$$
$$U - N = \{0, 1, 2, 3, 4, 5, 6, 7, 8, 9\} - \{6\} = \{0, 1, 2, 3, 4, 5, 7, 8, 9\}$$

The complement is closely related to the relative complement (see the preceding section). Both operations are similar to subtraction. The main difference is that the complement is *always* subtraction of a set from U, but the relative complement is subtraction of a set from any other set.

The symbol for the complement is ′, so you can write the following:

$M' = \{0, 2, 4, 6, 8\}$

$N' = \{0, 1, 2, 3, 4, 5, 7, 8, 9\}$

Part V

The X-Files: Introduction to Algebra

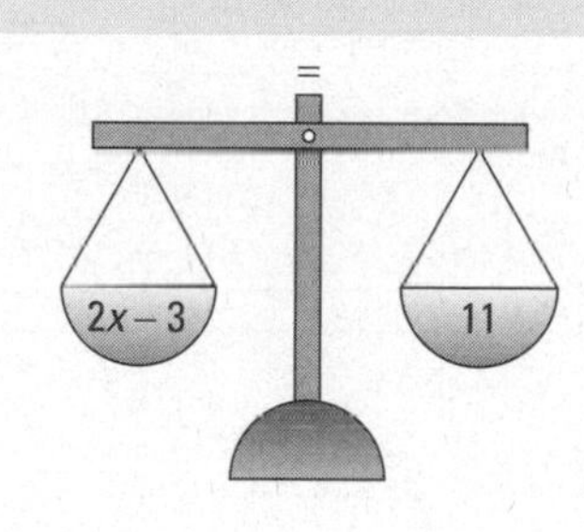

Expression	*Number of Terms*	*Terms*
$5x$	One	$5x$
$-5x + 2$	Two	$-5x$ and 2
$x^2y + \frac{z}{3} - xyz + 8$	Four	x^2y, $\frac{z}{3}$, xyz, and 8

Term	*Coefficient*	*Variable*
$-4x^3$	–4	x^3
x^2	1	x^2
$-x$	–1	x
–7	–7	None

For more info on simplifying algebraic expressions and to discover a trick that makes factoring quadratic polynomials easy, go to `www.dummies.com/extras/basicmathandprealgebra`.

In this part...

- Evaluate, simplify, and factor algebraic expressions
- Keep algebraic equations balanced and solve them by isolating the variable
- Use algebra to solve word problems too difficult to solve with just arithmetic

Chapter 21

Enter Mr. X: Algebra and Algebraic Expressions

In This Chapter

- Meeting Mr. X head-on
- Understanding how a variable such as x stands for a number
- Using substitution to evaluate an algebraic expression
- Identifying and rearranging the terms in any algebraic expression
- Simplifying algebraic expressions

You never forget your first love, your first car, or your first x. Unfortunately for some folks, remembering their first x in algebra is similar to remembering their first love who stood them up at the prom or their first car that broke down someplace in Mexico.

The most well-known fact about algebra is that it uses letters — like x — to represent numbers. So if you have a traumatic x-related tale, all I can say is that the future will be brighter than the past.

What good is algebra? That question is a common one, and it deserves a decent answer. Algebra is used for solving problems that are just too difficult for ordinary arithmetic. And because number crunching is so much a part of the modern world, algebra is everywhere (even if you don't see it): architecture, engineering, medicine, statistics, computers, business, chemistry, physics, biology, and, of course, higher math. Anywhere that numbers are useful, algebra is there. That fact is why virtually every college and university insists that you leave (or enter) with at least a passing familiarity with algebra.

In this chapter, I introduce (or reintroduce) you to that elusive little fellow, Mr. *X,* in a way that's bound to make him seem a little friendlier. Then I show you how *algebraic expressions* are similar to and different from the arithmetic expressions that you're used to working with. (For a refresher on arithmetic expressions, see Chapter 5.)

Seeing How X Marks the Spot

In math, x stands for a number — any number. Any letter that you use to stand for a number is a *variable,* which means that its value can *vary* — that is, its value is uncertain. In contrast, a number in algebra is often called a *constant* because its value is *fixed.*

Sometimes you have enough information to find out the identity of x. For example, consider the following:

$$2 + 2 = x$$

Obviously, in this equation, x stands for the number 4. But other times, what the number x stands for stays shrouded in mystery. For example:

$$x > 5$$

In this inequality, x stands for some number greater than 5 — maybe 6, maybe $7\frac{1}{2}$, maybe 542.002.

Expressing Yourself with Algebraic Expressions

In Chapter 5, I introduce you to arithmetic expressions: strings of numbers and operators that can be evaluated or placed on one side of an equation. For example:

$$2+3$$

$$7\times1.5-2$$

$$2^4-|-4|-\sqrt{400}$$

In this chapter, I introduce you to another type of mathematical expression: the algebraic expression. An *algebraic expression* is any string of mathematical symbols that can be placed on one side of an equation and that includes at least one variable.

Here are a few examples of algebraic expressions:

$$5x$$

$$-5x+2$$

$$x^2y-xy^2+\frac{z}{3}-xyz+1$$

As you can see, the difference between arithmetic and algebraic expressions is simply that an algebraic expression includes at least one variable.

In this section, I show you how to work with algebraic expressions. First, I demonstrate how to evaluate an algebraic expression by substituting the values of its variables. Then I show you how to separate an algebraic expression into one or more terms, and I walk through how to identify the coefficient and the variable part of each term.

Evaluating algebraic expressions

To evaluate an algebraic expression, you need to know the numerical value of every variable. For each variable in the expression, substitute, or plug in, the number that it stands for and then evaluate the expression.

In Chapter 5, I show you how to evaluate an arithmetic expression. Briefly, this means finding the value of that expression as a single number (flip to Chapter 5 for more on evaluating).

Knowing how to evaluate arithmetic expressions comes in handy for evaluating algebraic expressions. For example, suppose you want to evaluate the following expression:

$$4x - 7$$

Note that this expression contains the variable *x*, which is unknown, so the value of the whole expression is also unknown.

An algebraic expression can have any number of variables, but you usually don't work with expressions that have more than two or maybe three, at the most. You can use any letter as a variable, but *x*, *y*, and *z* tend to get a lot of mileage.

Suppose in this case that $x = 2$. To evaluate the expression, substitute 2 for *x* everywhere it appears in the expression:

$$4(2) - 7$$

After you make the substitution, you're left with an arithmetic expression, so you can finish your calculations to evaluate the expression:

$$= 8 - 7 = 1$$

So given $x = 2$, the algebraic expression $4x - 7 = 1$.

Now suppose you want to evaluate the following expression, where $x = 4$:

$$2x^2 - 5x - 15$$

Again, the first step is to substitute 4 for x everywhere this variable appears in the expression:

$$= 2(4^2) - 5(4) - 15$$

Now evaluate according to the order of operations explained in Chapter 5. You do powers first, so begin by evaluating the exponent 4^2, which equals 4×4:

$$= 2(16) - 5(4) - 15$$

Now proceed to the multiplication, moving from left to right:

$$= 32 - 5(4) - 15$$
$$= 32 - 20 - 15$$

Then evaluate the subtraction, again from left to right:

$$= 12 - 15$$
$$= -3$$

So given $x = 4$, the algebraic expression $2x^2 - 5x - 15 = -3$.

You aren't limited to expressions of only one variable when using substitution. As long as you know the value of every variable in the expression, you can evaluate algebraic expressions with any number of variables. For example, suppose you want to evaluate this expression:

$$3x^2 + 2xy - xyz$$

To evaluate it, you need the values of all three variables:

$$x = 3$$
$$y = -2$$
$$z = 5$$

The first step is to substitute the equivalent value for each of the three variables wherever you find them:

$$3(3^2) + 2(3)(-2) - (3)(-2)(5)$$

Now use the rules for order of operations from Chapter 5. Begin by evaluating the exponent 3^2:

$$= 3(9) + 2(3)(-2) - (3)(-2)(5)$$

Next, evaluate the multiplication from left to right (if you need to know more about the rules for multiplying negative numbers, check out Chapter 4):

$$= 27 + (-12) - (-30)$$

Now all that's left is addition and subtraction. Evaluate from left to right, remembering the rules for adding and subtracting negative numbers in Chapter 4:

$$= 15 - (-30) = 15 + 30 = 45$$

So given the three values for x, y, and z, the algebraic expression $3x^2 + 2xy - xyz = 45$.

For practice, copy this expression and the three values on a separate piece of paper, close the book, and see whether you can substitute and evaluate on your own to get the same answer.

Coming to algebraic terms

A *term* in an algebraic expression is any chunk of symbols set off from the rest of the expression by either addition or subtraction. As algebraic expressions get more complex, they begin to string themselves out in more terms. Here are some examples:

Expression	***Number of Terms***	***Terms***
$5x$	One	$5x$
$-5x + 2$	Two	$-5x$ and 2
$x^2y + \frac{z}{3} - xyz + 8$	Four	x^2y, $\frac{z}{3}$, $-xyz$, and 8

No matter how complicated an algebraic expression gets, you can always separate it out into one or more terms.

When separating an algebraic expression into terms, group the plus or minus sign with the term that it immediately precedes.

When a term has a variable, it's called an *algebraic term*. When it doesn't have a variable, it's called a *constant*. For example, look at the following expression:

$$x^2y + \frac{z}{3} - xyz + 8$$

The first three terms are algebraic terms, and the last term is a constant. As you can see, in algebra, *constant* is just a fancy word for *number*.

Terms are really useful to know about because you can follow rules to move them, combine them, and perform the Big Four operations on them. All these skills are important for solving equations, which I explain in the next chapter. But for now, this section explains a bit about terms and some of their traits.

Making the commute: Rearranging your terms

When you understand how to separate an algebraic expression into terms, you can go one step further by rearranging the terms in any order you like. Each term moves as a unit, kind of like a group of people carpooling to work together — everyone in the car stays together for the whole ride.

For example, suppose you begin with the expression $-5x + 2$. You can rearrange the two terms of this expression without changing its value. Notice that each term's sign stays with that term, although dropping the plus sign at the beginning of an expression is customary:

$$= 2 - 5x$$

Rearranging terms in this way doesn't affect the value of the expression because addition is *commutative* — that is, you can rearrange things that you're adding without changing the answer. (See Chapter 4 for more on the commutative property of addition.)

For example, suppose $x = 3$. Then the original expression and its rearrangement evaluate as follows (using the rules that I outline earlier in "Evaluating algebraic expressions"):

$$-5x + 2$$
$$= -5(3) + 2$$
$$= -15 + 2$$
$$= 13$$

$$2 - 5x$$
$$= 2 - 5(3)$$
$$= 2 - 15$$
$$= -13$$

Rearranging expressions in this way becomes handy later in this chapter, when you simplify algebraic expressions. As another example, suppose you have this expression:

$$4x - y + 6$$

You can rearrange it in a variety of ways:

$$= 6 + 4x - y$$
$$= -y + 4x + 6$$

Because the term $4x$ has no sign, it's positive, so you can write in a plus sign as needed when rearranging terms.

As long as each term's sign stays with that term, rearranging the terms in an expression has no effect on its value.

For example, suppose that $x = 2$ and $y = 3$. Here's how to evaluate the original expression and the two rearrangements:

$4x - y + 6$	$6 + 4x - y$	$-y + 4x + 6$
$= 4(2) - 3 + 6$	$= 6 + 4(2) - 3$	$= -3 + 4(2) + 6$
$= 8 - 3 + 6$	$= 6 + 8 - 3$	$= -3 + 8 + 6$
$= 5 + 6$	$= 14 - 3$	$= 5 + 6$
$= 11$	$= 11$	$= 11$

Identifying the coefficient and variable

Every term in an algebraic expression has a coefficient. The *coefficient* is the signed numerical part of a term in an algebraic expression — that is, the number and the sign (+ or –) that goes with that term. For example, suppose you're working with the following algebraic expression:

$$-4x^3 + x^2 - x - 7$$

The following table shows the four terms of this expression, with each term's coefficient:

Term	***Coefficient***	***Variable***
$-4x^3$	-4	x^3
x^2	1	x^2
$-x$	-1	x
-7	-7	none

Notice that the sign associated with the term is part of the coefficient. So the coefficient of $-4x^3$ is –4.

When a term appears to have no coefficient, the coefficient is actually 1. So the coefficient of x^2 is 1, and the coefficient of $-x$ is –1. And when a term is a constant (just a number), that number with its associated sign is the coefficient. So the coefficient of the term –7 is simply –7.

By the way, when the coefficient of any algebraic term is 0, the expression equals 0 no matter what the variable part looks like:

$0x = 0$ $\qquad 0xyz = 0$ $\qquad 0x^3y^4z^{10} = 0$

In contrast, the *variable part* of an expression is everything except the coefficient. The previous table shows the four terms of the same expression, with each term's variable part.

Identifying like terms

Like terms (or *similar terms*) are any two algebraic terms that have the same variable part — that is, both the letters and their exponents have to be exact matches. Here are some examples:

Variable Part	***Examples of Like Terms***		
x	$4x$	$12x$	$99.9x$
x^2	$6x^2$	$-20x^2$	$\frac{8}{3}x^2$
y	y	$1{,}000y$	πy
xy	$-7xy$	$800xy$	$\frac{22}{7}xy$
x^3y^3	$3x^3y^3$	$-111x^3y^3$	$3.14x^3y^3$

As you can see, in each example, the variable part in all three like terms is the same. Only the coefficient changes, and it can be any real number: positive or negative, whole number, fraction, or decimal — or even an irrational number such as π. (For more on real numbers, see Chapter 25.)

Considering algebraic terms and the Big Four

In this section, I get you up to speed on how to apply the Big Four to algebraic expressions. For now, just think of working with algebraic expressions as a set of tools that you're collecting, for use when you get on the job. You find how useful these tools are in Chapter 22, when you begin solving algebraic equations.

Adding terms

Add like terms by adding their coefficients and keeping the same variable part.

For example, suppose you have the expression $2x + 3x$. Remember that $2x$ is just shorthand for $x + x$, and $3x$ means simply $x + x + x$. So when you add them up, you get the following:

$$= x + x + x + x + x = 5x$$

As you can see, when the variable parts of two terms are the same, you add these terms by adding their coefficients: $2x + 3x = (2 + 3)x$. The idea here is roughly similar to the idea that 2 apples + 3 apples = 5 apples.

You *cannot* add nonlike terms. Here are some cases in which the variables or their exponents are different:

$$2x + 3y$$
$$2yz + 3y$$
$$2x^2 + 3x$$

In these cases, you can't simplify the expression. You're faced with a situation that's similar to 2 apples + 3 oranges. Because the units (apples and oranges) are different, you can't combine terms. (See Chapter 4 for more on how to work with units.)

Subtracting terms

Subtraction works much the same as addition. Subtract like terms by finding the difference between their coefficients and keeping the same variable part.

For example, suppose you have $3x - x$. Recall that $3x$ is simply shorthand for $x + x + x$. So doing this subtraction gives you the following:

$$x + x + x - x = 2x$$

No big surprises here. You simply find $(3 - 1)x$. This time, the idea roughly parallels the idea that \$3 – \$1 = \$2.

Here's another example:

$$2x - 5x$$

Again, no problem, as long as you know how to work with negative numbers (see Chapter 4 if you need details). Just find the difference between the coefficients:

$$= (2 - 5)x = -3x$$

In this case, recall that \$2 – \$5 = –\$3 (that is, a debt of \$3).

You *cannot* subtract nonlike terms. For example, you can't subtract either of the following:

$$7x - 4y$$

$$7x^2y - 4xy^2$$

As with addition, you can't do subtraction with different variables. Think of this as trying to figure out \$7 – 4 pesos. Because the units in this case (dollars versus pesos) are different, you're stuck. (See Chapter 4 for more on working with units.)

Multiplying terms

Unlike adding and subtracting, you can multiply nonlike terms. Multiply *any* two terms by multiplying their coefficients and combining — that is, by collecting or gathering up — all the variables in each term into a single term, as I show you next.

For example, suppose you want to multiply $5x(3y)$. To get the coefficient, multiply 5×3. To get the algebraic part, combine the variables x and y:

$$= 5(3)xy = 15xy$$

Now suppose you want to multiply $2x(7x)$. Again, multiply the coefficients and collect the variables into a single term:

$$= 7(2)xx = 14xx$$

Remember that x^2 is shorthand for xx, so you can write the answer more efficiently:

$$=14x^2$$

Here's another example. Multiply all three coefficients together and gather up the variables:

$$2x^2(3y)(4xy)$$
$$= 2(3)(4)x^2xyy$$
$$= 24x^3y^2$$

As you can see, the exponent 3 that's associated with x is just the count of how many x's appear in the problem. The same is true of the exponent 2 associated with y.

A fast way to multiply variables with exponents is to add the exponents together. For example:

$$(x^4y^3)(x^2y^5)(x^6y) = x^{12}y^9$$

In this example, I added the exponents of the x's ($4 + 2 + 6 = 12$) to get the exponent of x in the expression. Similarly, I added the exponents of the y's ($3 + 5 + 1 = 9$ — don't forget that $y = y^1$!) to get the exponent of y in the expression.

Dividing terms

It's customary to represent division of algebraic expressions as a fraction instead of using the division sign (÷). So division of algebraic terms really looks like reducing a fraction to lowest terms (see Chapter 9 for more on reducing).

To divide one algebraic term by another, follow these steps:

1. **Make a fraction of the two terms.**

   Suppose you want to divide $3xy$ by $12x^2$. Begin by turning the problem into a fraction:

   $$\frac{3xy}{12x^2}$$

2. **Cancel out factors in coefficients that are in both the numerator and the denominator.**

   In this case, you can cancel out a 3. Notice that when the coefficient in xy becomes 1, you can drop it:

   $$= \frac{xy}{4x^2}$$

3. Cancel out any variable that's in both the numerator and the denominator.

You can break x^2 out as xx:

$$= \frac{xy}{4xx}$$

Now you can clearly cancel an x in both the numerator and the denominator:

$$= \frac{y}{4x}$$

As you can see, the resulting fraction is really a reduced form of the original.

As another example, suppose you want to divide $-6x^2yz^3$ by $-8x^2y^2z$. Begin by writing the division as a fraction:

$$\frac{-6x^2yz^3}{-8x^2y^2z}$$

First, reduce the coefficients. Notice that, because both coefficients were originally negative, you can cancel out both minus signs as well:

$$= \frac{3x^2yz^3}{4x^2y^2z}$$

Now you can begin canceling variables. I do this in two steps, as before:

$$= \frac{3xxyzzz}{4xxyyz}$$

At this point, just cross out any occurrence of a variable that appears in both the numerator and the denominator:

$$= \frac{3zz}{4y}$$

$$= \frac{3z^2}{4y}$$

You can't cancel out variables or coefficients if either the numerator or the denominator has more than one term in it. This is a very common mistake in algebra, so don't let it happen to you!

Simplifying Algebraic Expressions

As algebraic expressions grow more complex, simplifying them can make them easier to work with. Simplifying an expression means (quite simply!) making it smaller and easier to manage. You see how important simplifying expressions becomes when you begin solving algebraic equations.

For now, think of this section as a kind of algebra toolkit. Here I show you *how* to use these tools. In Chapter 22, I show you *when* to use them.

Combining like terms

When two algebraic terms contain like terms (when their variables match), you can add or subtract them (see the earlier section "Considering algebraic terms and the Big Four"). This feature comes in handy when you're trying to simplify an expression. For example, suppose you're working with the following expression:

$$4x - 3y + 2x + y - x + 2y$$

As it stands, this expression has six terms. But three terms have the variable x and the other three have the variable y. Begin by rearranging the expression so that all like terms are grouped together:

$$= 4x + 2x - x - 3y + y + 2y$$

Now you can add and subtract like terms. I do this in two steps, first for the x terms and then for the y terms:

$$= 5x - 3y + y + 2y$$
$$= 5x + 0y$$
$$= 5x$$

Notice that the x terms simplify to $5x$, and the y terms simplify to $0y$, which is 0, so the y terms drop out of the expression altogether.

Here's a somewhat more complicated example that has variables with exponents:

$$12x - xy - 3x^2 + 8y + 10xy + 3x^2 - 7x$$

This time, you have four different types of terms. As a first step, you can rearrange these terms so that groups of like terms are all together (I underline these four groups so you can see them clearly):

$$= \underline{12x - 7x} \; \underline{-xy + 10xy} \; \underline{-3x^2 + 3x^2} \; \underline{+8y}$$

Now combine each set of like terms:

$$= 5x + 9xy + 0x^2 + 8y$$

This time, the x^2 terms add up to 0, so they drop out of the expression altogether:

$$= 5x + 9xy + 8y$$

Removing parentheses from an algebraic expression

Parentheses keep parts of an expression together as a single unit. In Chapter 5, I show you how to handle parentheses in an arithmetic expression. This skill is also useful with algebraic expressions. As you find when you begin solving algebraic equations in Chapter 22, getting rid of parentheses is often the first step toward solving a problem. In this section, I show how to handle the Big Four operations with ease.

Drop everything: Parentheses with a plus sign

When an expression contains parentheses that come right after a plus sign (+), you can just remove the parentheses. Here's an example:

$$2x \underline{+(3x - y)} + 5y$$
$$= 2x \underline{+3x - y} + 5y$$

Now you can simplify the expression by combining like terms:

$$= 5x + 4y$$

When the first term inside the parentheses is negative, when you drop the parentheses, the minus sign replaces the plus sign. For example:

$$6x+(-2x+y)-4y$$
$$=6x-2x+y-4y$$
$$=4x-3x$$

Sign turnabout: Parentheses with a minus sign

Sometimes an expression contains parentheses that come right after a minus sign (–). In this case, change the sign of every term inside the parentheses to the opposite sign; then remove the parentheses.

Consider this example:

$$6x - (2xy - 3y) + 5xy$$

A minus sign is in front of the parentheses, so you need to change the signs of both terms in the parentheses and remove the parentheses. Notice that the term $2xy$ appears to have no sign because it's the first term inside the parentheses. This expression really means the following:

$$= 6x \underline{- (+2xy - 3y)} + 5xy$$

You can see how to change the signs:

$$= 6x \underline{- 2xy + 3y} + 5xy$$

At this point, you can combine the two xy terms:

$$= 6x + 3xy + 3y$$

Distribution: Parentheses with no sign

When you see nothing between a number and a set of parentheses, it means multiplication. For example,

$$2(3) = 6 \quad 4(4) = 16 \quad 10(15) = 150$$

This notation becomes much more common with algebraic expressions, replacing the multiplication sign (×) to avoid confusion with the variable x:

$$3(4x) = 12x \quad 4x(2x) = 8x^2 \quad 3x(7y) = 21xy$$

To remove parentheses without a sign, multiply the term outside the parentheses by every term inside the parentheses; then remove the parentheses. When you follow those steps, you're using the *distributive property*.

Here's an example:

$$2(3x - 5y + 4)$$

In this case, multiply 2 by each of the three terms inside the parentheses:

$$= 2(3x) + 2(-5y) + 2(4)$$

For the moment, this expression looks more complex than the original one, but now you can get rid of all three sets of parentheses by multiplying:

$$= 6x - 10y + 8$$

Multiplying by every term inside the parentheses is simply distribution of multiplication over addition — also called the *distributive property* — which I discuss in Chapter 4.

As another example, suppose you have the following expression:

$$-2x(-3x + y + 6) + 2xy - 5x^2$$

Begin by multiplying $-2x$ by the three terms inside the parentheses:

$$= -2x(-3x) - 2x(y) - 2x(6) + 2xy - 5x^2$$

The expression looks worse than when you started, but you can get rid of all the parentheses by multiplying:

$$= 6x^2 - 2xy - 12x + 2xy - 5x^2$$

Now you can combine like terms:

$$= x^2 - 12x$$

Parentheses by FOILing

Sometimes, expressions have two sets of parentheses next to each other without a sign between them. In that case, you need to multiply *every term* inside the first set by *every term* inside the second.

When you have two terms inside each set of parentheses, you can use a process called FOILing. This is really just the distributive property, as I show you below. The word *FOIL* is an acronym to help you make sure you multiply the correct terms. It stands for *F*irst, *O*utside, *I*nside, and *L*ast.

Here's how the process works. In this example, you're simplifying the expression $(2x - 2)(3x - 6)$:

1. **Start out by multiplying the two First terms in the parentheses.**

   The first term in the first set of parentheses is $2x$, and $3x$ is the first term in the second set of parentheses: $(\underline{2x} - 2)(\underline{3x} - 6)$.

   F: Multiply the first terms: $2x(3x) = 6x^2$

2. **Multiply the two *Outside* terms.**

   The two outside terms, $2x$ and -6, are on the ends: $(\underline{2x} - 2)(3x \underline{- 6})$

   O: Multiply the outside terms: $2x(-6) = -12x$

3. **Multiply the two *Inside* terms.**

   The two terms in the middle are -2 and $3x$: $(2x \underline{- 2})(\underline{3x} - 6)$

   I: Multiply the middle terms: $-2(3x) = -6x$

4. **Multiply the two *Last* terms.**

   The last term in the first set of parentheses is -2, and -6 is the last term in the second set: $(2x \underline{- 2})(3x \underline{- 6})$

   L: Multiply the last terms: $-2(-6) = 12$

Add these four results together to get the simplified expression:

$$(2x - 2)(3x - 6) = 6x^2 - 6x - 12x + 12$$

In this case, you can simplify this expression still further by combining the like terms $-12x$ and $-6x$:

$$= 6x^2 - 18x + 12$$

Notice that, during this process, you multiply every term inside one set of parentheses by every term inside the other set. FOILing just helps you keep track and make sure you've multiplied everything.

FOILing is really just an application of the distributive property, which I discuss in the section preceding this one. In other words, $(2x - 2)(3x - 6)$ is really the same as $2x(3x - 6) + -2(3x - 6)$ when distributed. Then distributing again gives you $6x^2 - 6x - 12x + 12$.

Chapter 22

Unmasking Mr. X: Algebraic Equations

In This Chapter

- Using variables (such as x) in equations
- Knowing some quick ways to solve for x in simple equations
- Understanding the *balance scale method* for solving equations
- Rearranging terms in an algebraic equation
- Isolating algebraic terms on one side of an equation
- Removing parentheses from an equation
- Cross-multiplying to remove fractions

When it comes to algebra, solving equations is the main event.

Solving an algebraic equation means finding out what number the variable (usually x) stands for. Not surprisingly, this process is called *solving for x,* and when you know how to do it, your confidence — not to mention your grades in your algebra class — will soar through the roof.

This chapter is all about solving for x. First, I show you a few informal methods to solve for x when an equation isn't too difficult. Then I show you how to solve more difficult equations by thinking of them as a balance scale.

The balance scale method is really the heart of algebra (yes, algebra has a heart, after all!). When you understand this simple idea, you're ready to solve more complicated equations, using all the tools I show you in Chapter 21, such as simplifying expressions and removing parentheses. You find out how to extend these skills to algebraic equations. Finally, I show you how cross-multiplying (see Chapter 9) can make solving algebraic equations with fractions a piece of cake.

By the end of this chapter, you'll have a solid grasp of a bunch of ways to solve equations for the elusive and mysterious x.

Understanding Algebraic Equations

An algebraic equation is an equation that includes at least one variable — that is, a letter (such as x) that stands for a number. *Solving* an algebraic equation means finding out what number x stands for.

In this section, I show you the basics of how a variable like x works its way into an equation in the first place. Then I show you a few quick ways to *solve for* x when an equation isn't too difficult.

Using x in equations

As you discover in Chapter 5, an *equation* is a mathematical statement that contains an equals sign. For example, here's a perfectly good equation:

$$7 \times 9 = 63$$

At its heart, a variable (such as x) is nothing more than a placeholder for a number. You're probably used to equations that use other placeholders: One number is purposely left as a blank or replaced by an underline or a question mark, and you're supposed to fill it in. Usually, this number comes after the equals sign. For example:

$$8 + 2 =$$
$$12 - 3 = \_\_$$
$$14 \div 7 = ?$$

As soon as you're comfortable with addition, subtraction, or whatever, you can switch the equation around a bit:

$$9 + \_\_ = 14$$
$$? \times 6 = 18$$

When you stop using underlines and question marks and start using variables such as x to stand for the part of the equation you want to figure out, bingo! You have an algebra problem:

$$4 + 1 = x$$
$$12 \div x = 3$$
$$x - 13 = 30$$

Choosing among four ways to solve algebraic equations

You don't need to call an exterminator just to kill a bug. Similarly, algebra is strong stuff, and you don't always need it to solve an algebraic equation.

Generally, you have four ways to solve algebraic equations such as the ones I introduce earlier in this chapter. In this section, I introduce them in order of difficulty.

Eyeballing easy equations

You can solve easy problems just by looking at them. For example:

$$5 + x = 6$$

When you look at this problem, you can see that $x = 1$. When a problem is this easy and you can see the answer, you don't need to go to any particular trouble to solve it.

Rearranging slightly harder equations

When you can't see an answer just by looking at a problem, sometimes rearranging the problem helps to turn it into one that you can solve using a Big Four operation. For example:

$$6x = 96$$

You can rearrange this problem using inverse operations, as I show you in Chapter 4, changing multiplication to division:

$$x = \frac{96}{6}$$

Now solve the problem by division (long or otherwise) to find that $x = 16$.

Guessing and checking equations

You can solve some equations by guessing an answer and then checking to see whether you're right. For example, suppose you want to solve the following equation:

$$3x + 7 = 19$$

To find out what x equals, start by guessing that $x = 2$. Now check to see whether you're right by substituting 2 for x in the equation:

$3(2) + 7 = 13$ WRONG! (13 is less than 19.)

$3(5) + 7 = 22$ 19 WRONG! (22 is greater than 19.)

$3(4) + 7 = 19$ RIGHT!

With only three guesses, you found that $x = 4$.

Applying algebra to more difficult equations

When an algebraic equation gets hard enough, you find that looking at it and rearranging it just isn't enough to solve it. For example:

$$11x - 13 = 9x + 3$$

You probably can't tell what x equals just by looking at this problem. You also can't solve it just by rearranging it, using an inverse operation. And guessing and checking would be very tedious. Here's where algebra comes into play.

Algebra is especially useful because you can follow mathematical rules to find your answer. Throughout the rest of this chapter, I show you how to use the rules of algebra to turn tough problems like this one into problems that you can solve.

The Balancing Act: Solving for x

As I show you in the preceding section, some problems are too complicated to find out what the variable (usually x) equals just by eyeballing it or rearranging it. For these problems, you need a reliable method for getting the right answer. I call this method the *balance scale.*

The balance scale allows you to *solve for x* — that is, find the number that x stands for — in a step-by-step process that always works. In this section, I show you how to use the balance scale method to solve algebraic equations.

Striking a balance

The equals sign in any equation means that both sides balance. To keep that equals sign, you have to maintain that balance. In other words, whatever you do to one side of an equation, you have to do to the other.

For example, here's a balanced equation:

$$\frac{1 + 2 = 3}{\Delta}$$

Illustration by Wiley, Composition Services Graphics

If you add 1 to one side of the equation, the scale goes out of balance.

$$\frac{1 + 2 + 1 \neq 3}{\Delta}$$

Illustration by Wiley, Composition Services Graphics

But if you add 1 to *both* sides of the equation, the scale stays balanced:

$$\frac{1 + 2 + 1 = 3 + 1}{\Delta}$$

Illustration by Wiley, Composition Services Graphics

You can add any number to the equation, as long as you do it to both sides. And in math, *any number* means x:

$$1 + 2 + x = 3 + x$$

Remember that x is the same wherever it appears in a single equation or problem.

This idea of changing both sides of an equation equally isn't limited to addition. You can just as easily subtract an x, or even multiply or divide by x, as long as you do the same to both sides of the equation:

$$\begin{aligned} \text{Subtract: } & 1+2-x=3-x \\ \text{Multiply: } & (1+2)x=3x \\ \text{Divide: } & \frac{1+2}{x}=\frac{3}{x} \end{aligned}$$

Using the balance scale to isolate x

The simple idea of balance is at the heart of algebra, and it enables you to find out what x is in many equations. When you solve an algebraic equation, the goal is to *isolate* x — that is, to get x alone on one side of the equation and some number on the other side. In algebraic equations of middling difficulty, this is a three-step process:

1. **Get all constants (non-x terms) on one side of the equation.**
2. **Get all x-terms on the other side of the equation.**
3. **Divide to isolate x.**

For example, take a look at the following problem:

$$11x - 13 = 9x + 3$$

As you follow the steps, notice how I keep the equation balanced at each step:

1. **Get all the constants on one side of the equation by adding 13 to both sides of the equation:**

$$\begin{array}{r} 11x - 13 = 9x + 3 \\ +13 = \quad +13 \\ \hline 11x \quad = 9x + 16 \end{array}$$

Because you've obeyed the rules of the balance scale, you know that this new equation is also correct. Now the only non-x term (16) is on the right side of the equation.

2. **Get all the x-terms on the other side by subtracting $9x$ from both sides of the equation:**

$$\begin{array}{r} 11x = 9x + 16 \\ -9x \; -9x \quad \\ \hline 2x = \quad 16 \end{array}$$

Again, the balance is preserved, so the new equation is correct.

3. **Divide by 2 to isolate x:**

$$\frac{2x}{2} = \frac{16}{2}$$
$$x = 8$$

To check this answer, you can simply substitute 8 for x in the original equation:

$$11(8) - 13 = 9(8) + 3$$
$$88 - 13 = 72 + 3$$
$$75 = 75 \checkmark$$

This checks out, so 8 is the correct value of x.

Rearranging Equations and Isolating x

When you understand how algebra works like a balance scale, as I show you in the preceding section, you can begin to solve more-difficult algebraic equations. The basic tactic is always the same: Changing both sides of the equation equally at every step, try to isolate x on one side of the equation.

In this section, I show you how to put your skills from Chapter 21 to work solving equations. First, I show you how rearranging the terms in an expression is similar to rearranging them in an algebraic equation. Next, I show you how removing parentheses from an equation can help you solve it. Finally, you discover how cross-multiplication is useful for solving algebraic equations with fractions.

Rearranging terms on one side of an equation

Rearranging terms becomes all-important when working with equations. For example, suppose you're working with this equation:

$$5x - 4 = 2x + 2$$

When you think about it, this equation is really two expressions connected with an equals sign. And of course, that's true of *every* equation. That's why everything you find out about expressions in Chapter 21 is useful for solving equations. For example, you can rearrange the terms on one side of an equation. So here's another way to write the same equation:

$$-4 + 5x = 2x + 2$$

And here's a third way:

$$-4 + 5x = 2 + 2x$$

This flexibility to rearrange terms comes in handy when you're solving equations.

Moving terms to the other side of the equals sign

Earlier in this chapter, I show you how an equation is similar to a balance scale. For example, take a look at Figure 22-1.

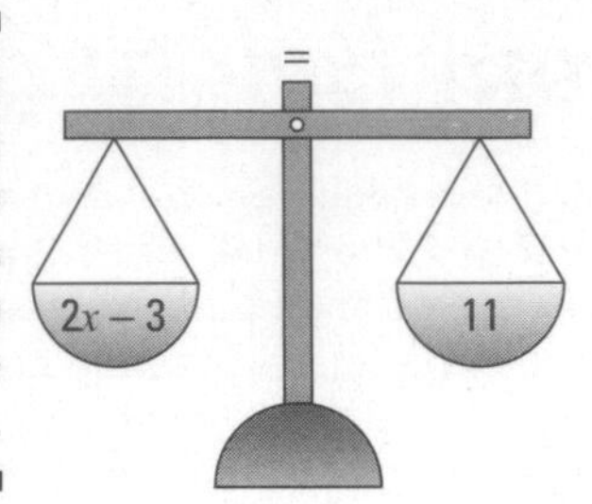

Figure 22-1: Showing how an equation is similar to a balance scale.

Illustration by Wiley, Composition Services Graphics

To keep the scale balanced, if you add or remove anything on one side, you must do the same on the other side. For example:

$$\begin{array}{r} 2x-3=11 \\ \underline{-2x \qquad\quad -2x} \\ -3=11-2x \end{array}$$

Now take a look at these two versions of this equation side by side:

$$2x - 3 = 11 \qquad -3 = 11 - 2x$$

In the first version, the term $2x$ is on the left side of the equals sign. In the second, the term $-2x$ is on the right side. This example illustrates an important rule.

When you move any term in an expression to the other side of the equals sign, change its sign (from plus to minus or from minus to plus).

As another example, suppose you're working with this equation:

$$4x - 2 = 3x + 1$$

You have x's on both sides of the equation, so say you want to move the $3x$. When you move the term $3x$ from the right side to the left side, you have to change its sign from plus to minus (technically, you're subtracting $3x$ from both sides of the equation).

$$4x - 2 - 3x = 1$$

After that, you can simplify the expression on the left side of the equation by combining like terms:

$$x - 2 = 1$$

At this point, you can probably see that $x = 3$ because $3 - 2 = 1$. But just to be sure, move the -2 term to the right side and change its sign:

$$x = 1 + 2$$
$$x = 3$$

To check this result, substitute a 3 wherever x appears in the original equation:

$$4x - 2 = 3x + 1$$
$$4(3) - 2 = 3(3) + 1$$
$$12 - 2 = 9 + 1$$
$$10 = 10 \checkmark$$

As you can see, moving terms from one side of an equation to the other can be a big help when you're solving equations.

Removing parentheses from equations

Chapter 21 gives you a treasure trove of tricks for simplifying expressions, and they come in handy when you're solving equations. One key skill from that chapter is removing parentheses from expressions. This tactic is also indispensable when you're solving equations.

For example, suppose you have the following equation:

$$5x + (6x - 15) = 30 - (x - 7) + 8$$

Your mission is to get all the x terms on one side of the equation and all the constants on the other. As the equation stands, however, x terms and constants are "locked together" inside parentheses. In other words, you can't isolate the x terms from the constants. So before you can isolate terms, you need to remove the parentheses from the equation.

Recall that an equation is really just two expressions connected by an equals sign. So you can start working with the expression on the left side. In this expression, the parentheses begin with a plus sign (+), so you can just remove them:

$$5x + \underline{6x - 15} = 30 - (x - 7) + 8$$

Now move on to the expression on the right side. This time, the parentheses come right after a minus sign (–). To remove them, change the sign of both terms inside the parentheses: x becomes $-x$, and -7 becomes 7:

$$5x + 6x - 15 = 30 \underline{- x + 7} + 8$$

Bravo! Now you can isolate x terms to your heart's content. Move the $-x$ from the right side to the left, changing it to x:

$$5x + 6x - 15 \underline{+x} = 30 + 7 + 8$$

Next, move –15 from the left side to the right, changing it to 15:

$$5x + 6x + x = 30 + 7 + 8 \underline{+15}$$

Now combine like terms on both sides of the equation:

$$12x = 30 + 7 + 8 + 15$$
$$12x = 60$$

Finally, get rid of the coefficient 12 by dividing:

$$\frac{12x}{12} = \frac{60}{12}$$
$$x = 5$$

As usual, you can check your answer by substituting 5 into the original equation wherever x appears:

$$5x + (6x - 15) = 30 - (x - 7) + 8$$
$$5(5) + [6(5) - 15) = 30 - (5 - 7) + 8$$
$$25 + (30 - 15) = 30 - (-2) + 8$$
$$25 + 15 = 30 + 2 + 8$$
$$40 = 40 \checkmark$$

Here's one more example:

$$11 + 3(-3x + 1) = 25 - (7x - 3) - 12$$

As in the preceding example, start out by removing both sets of parentheses. This time, however, on the left side of the equation, you have no sign between 3 and $(-3x + 1)$. But again, you can put your skills from Chapter 21 to use. To remove the parentheses, multiply 3 by both terms inside the parentheses:

$$11 - 9x + 3 = 25 - (7x - 3) - 12$$

On the right side, the parentheses begin with a minus sign, so remove the parentheses by changing all the signs inside the parentheses:

$$11 - 9x + 3 = 25 - 7x + 3 - 12$$

Now you're ready to isolate the x terms. I do this in one step, but take as many steps as you want:

$$-9x + 7x = 25 + 3 - 12 - 11 - 3$$

At this point, you can combine like terms:

$$-2x = 2$$

To finish, divide both sides by –2:

$$x = -1$$

Copy this example, and work through it a few times with the book closed.

Cross-multiplying

In algebra, cross-multiplication helps to simplify equations by removing unwanted fractions (and, honestly, when are fractions ever wanted?). As I discuss in Chapter 9, you can use cross-multiplication to find out whether two fractions are equal. You can use this same idea to solve algebra equations with fractions, like this one:

$$\frac{x}{2x-2} = \frac{2x+3}{4x}$$

This equation looks hairy. You can't do the division or cancel anything out because the fraction on the left has two terms in the denominator, and the fraction on the right has two terms in the numerator (see Chapter 21 for info on dividing algebraic terms). However, an important piece of information that you have is that the fraction equals the fraction. So if you cross-multiply these two fractions, you get two results that are also equal:

$$x(4x) = (2x + 3)(2x - 2)$$

At this point, you have something you know how to work with. The left side is easy:

$$4x^2 = (2x+3)(2x-2)$$

The right side requires a bit of FOILing (flip to Chapter 21 for details):

$$4x^2 = 4x^2 - 4x + 6x - 6$$

Now all the parentheses are gone, so you can isolate the x terms. Because most of these terms are already on the right side of the equation, isolate them on that side:

$$6 = 4x^2 - 4x + 6x - 4x^2$$

Combining like terms gives you a pleasant surprise:

$$6 = 2x$$

The two x^2 terms cancel each other out. You may be able to eyeball the correct answer, but here's how to finish:

$$\frac{6}{2} = \frac{2x}{2}$$
$$3 = x$$

To check your answer, substitute 3 back into the original equation:

$$\frac{x}{2x-2} = \frac{2x+3}{4x}$$
$$\frac{3}{2(3)-2} = \frac{2(3)+3}{4(3)}$$
$$\frac{3}{6-2} = \frac{6+3}{12}$$
$$\frac{3}{4} = \frac{3}{4} \checkmark$$

So the answer $x = 3$ is correct.

Chapter 23

Putting Mr. X to Work: Algebra Word Problems

In This Chapter

- Solving algebra word problems in simple steps
- Choosing variables
- Using charts

Word problems that require algebra are among the toughest problems that students face — and the most common. Teachers just love algebra word problems because they bring together a lot of what you know, such as solving algebra equations (Chapters 21 and 22) and turning words into numbers (see Chapters 6, 13, and 18). And standardized tests virtually always include these types of problems.

In this chapter, I show you a five-step method for using algebra to solve word problems. Then I give you a bunch of examples that take you through all five steps.

Along the way, I give you some important tips that can make solving word problems easier. First, I show you how to choose a variable that makes your equation as simple as possible. Next, I give you practice organizing information from the problem into a chart. By the end of this chapter, you'll have a solid understanding of how to solve a wide variety of algebra word problems.

Solving Algebra Word Problems in Five Steps

Everything from Chapters 21 and 22 comes into play when you use algebra to solve word problems, so if you feel a little shaky on solving algebraic equations, flip back to those chapters for some review.

Throughout this section, I use the following word problem as an example:

> In three days, Alexandra sold a total of 31 tickets to her school play. On Tuesday, she sold twice as many tickets as on Wednesday. And on Thursday, she sold exactly 7 tickets. How many tickets did Alexandra sell on each day, Tuesday through Thursday?

Organizing the information in an algebra word problem by using a chart or picture is usually helpful. Here's what I came up with:

Tuesday:	Twice as many as on Wednesday
Wednesday:	?
Thursday:	7
Total:	31

At this point, all the information is in the chart, but the answer still may not be jumping out at you. In this section, I outline a step-by-step method that enables you to solve this problem — and much harder ones as well.

Here are the five steps for solving most algebra word problems:

1. **Declare a variable.**
2. **Set up the equation.**
3. **Solve the equation.**
4. **Answer the question that the problem asks.**
5. **Check your answer.**

Declaring a variable

As you know from Chapter 21, a variable is a letter that stands for a number. Most of the time, you don't find the variable x (or any other variable, for that matter) in a word problem. That omission doesn't mean you don't need algebra to solve the problem. It just means that you're going to have to put x into the problem yourself and decide what it stands for.

When you *declare a variable,* you say what that variable means in the problem you're solving.

Here are some examples of variable declarations:

Let m = the number of dead mice that the cat dragged into the house.

Let p = the number of times Marianne's husband promised to take out the garbage.

Let c = the number of complaints Arnold received after he painted his garage door purple.

In each case, you take a variable (m, p, or c) and give it a meaning by attaching it to a number.

Notice that the earlier chart for the sample problem has a big question mark next to *Wednesday.* This question mark stands for *some number,* so you may want to declare a variable that stands for this number. Here's how you do it:

Let w = the number of tickets that Alexandra sold on Wednesday.

Whenever possible, choose a variable with the same initial as what the variable stands for. This practice makes remembering what the variable means a lot easier, which will help you later in the problem.

For the rest of the problem, every time you see the variable w, keep in mind that it stands for the number of tickets that Alexandra sold on Wednesday.

Setting up the equation

After you have a variable to work with, you can go through the problem again and find other ways to use this variable. For example, Alexandra sold twice as many tickets on Tuesday as on Wednesday, so she sold $2w$ tickets on Tuesday. Now you have a lot more information to fill in on the chart:

Tuesday:	Twice as many as on Wednesday	$2w$
Wednesday:	?	w
Thursday:	7	7
Total:	31	31

You know that the total number of tickets, or the sum of the tickets she sold on Tuesday, Wednesday, and Thursday, is 31. With the chart filled in like that, you're ready to set up an equation to solve the problem:

$$2w + w + 7 = 31$$

Solving the equation

After you set up an equation, you can use the tricks from Chapter 22 to solve the equation for w. Here's the equation one more time:

$$2w + w + 7 = 31$$

For starters, remember that $2w$ really means $w + w$. So on the left, you know you really have $w + w + w$, or $3w$; you can simplify the equation a little bit, as follows:

$$3w + 7 = 31$$

The goal at this point is to try to get all the terms with w on one side of the equation and all the terms without w on the other side. So on the left side of the equation, you want to get rid of the 7. The inverse of addition is subtraction, so subtract 7 from both sides:

$$\begin{array}{l} 3w + 7 = 31 \\ \underline{\quad\ -7\ \ -7} \\ 3w \qquad = 24 \end{array}$$

You now want to isolate w on the left side of the equation. To do this, you have to undo the multiplication by 3, so divide both sides by 3:

$$\frac{3w}{3} = \frac{24}{3}$$
$$w = 8$$

Answering the question

You may be tempted to think that, after you've solved the equation, you're done. But you still have a bit more work to do. Look back at the problem, and you see that it asks you this question:

> How many tickets did Alexandra sell on each day, Tuesday through Thursday?

At this point, you have some information that can help you solve the problem. The problem tells you that Alexandra sold 7 tickets on Thursday. And because $w = 8$, you now know that she sold 8 tickets on Wednesday. And on Tuesday, she sold twice as many on Wednesday, so she sold 16. So Alexandra sold 16 tickets on Tuesday, 8 on Wednesday, and 7 on Thursday.

Checking your work

To check your work, compare your answer to the problem, line by line, to make sure every statement in the problem is true:

> In three days, Alexandra sold a total of 31 tickets to her school play.

That part is correct because $16 + 8 + 7 = 31$.

> On Tuesday, she sold twice as many tickets as on Wednesday.

Correct, because she sold 16 tickets on Tuesday and 8 on Wednesday.

> And on Thursday, she sold exactly 7 tickets.

Yep, that's right, too, so you're good to go.

Choosing Your Variable Wisely

Declaring a variable is simple, as I show you earlier in this chapter, but you can make the rest of your work a lot easier when you know how to choose your variable wisely. Whenever possible, choose a variable so that the equation you have to solve has no fractions, which are much more difficult to work with than whole numbers.

For example, suppose you're trying to solve this problem:

> Irina has three times as many clients as Toby. If they have 52 clients all together, how many clients does each person have?

The key sentence in the problem is "Irina has *three times as many* clients as Toby." It's significant because it indicates a relationship between Irina and Toby that's based on either *multiplication or division.* And to avoid fractions, you want to avoid division wherever possible.

Whenever you see a sentence that indicates you need to use either multiplication or division, choose your variable to represent the *smaller* number. In this case, Toby has fewer clients than Irina, so choosing t as your variable is the smart move.

Suppose you begin by declaring your variable as follows:

> Let t = the number of clients that Toby has.

Then, using that variable, you can make this chart:

Irina $3t$
Toby t

No fraction! To solve this problem, set up this equation:

Irina + Toby = 52

Plug in the values from the chart:

$$3t + t = 52$$

Now you can solve the problem easily, using what I show you in Chapter 22:

$$4t = 52$$
$$t = 13$$

Toby has 13 clients, so Irina has 39. To check this result — which I recommend highly earlier in this chapter! — note that 13 + 39 = 52.

Now suppose that, instead, you take the opposite route and decide to declare a variable as follows:

Let i = the number of clients that Irina has.

Given that variable, you have to represent Toby's clients using the fraction $\frac{i}{3}$, which leads to the same answer but a *lot* more work.

Solving More-Complex Algebraic Problems

Algebra word problems become more complex when the number of people or things you need to find out increases. In this section, the complexity increases to four and then five people. When you're done, you should feel comfortable solving algebra word problems of significant difficulty.

Charting four people

As in the previous section, a chart can help you organize information so you don't get confused. Here's a problem that involves four people:

> Alison, Jeremy, Liz, and Raymond participated in a canned goods drive at work. Liz donated three times as many cans as Jeremy, Alison donated

twice as many as Jeremy, and Raymond donated 7 more than Liz. Together the two women donated two more cans than the two men. How many cans did the four people donate altogether?

The first step, as always, is declaring a variable. Remember that, to avoid fractions, you want to declare a variable based on the person who brought in the fewest cans. Liz donated more cans than Jeremy, and so did Alison. Furthermore, Raymond donated more cans than Liz. So because Jeremy donated the fewest cans, declare your variable as follows:

Let j = the number of cans that Jeremy donated.

Now you can set up your chart as follows:

Jeremy	j
Liz	$3j$
Alison	$2j$
Raymond	Liz + 7 = $3j + 7$

This setup looks good because, as expected, there are no fractional amounts in the chart. The next sentence tells you that the women donated two more cans than the men, so make a word problem, as I show you in Chapter 6:

Liz + Alison = Jeremy + Raymond + 2

You can now substitute into this equation as follows:

$$3j + 2j = j + 3j + 7 + 2$$

With your equation set up, you're ready to solve. First, isolate the algebraic terms:

$$3j + 2j - j - 3j = 7 + 2$$

Combine like terms:

$$j = 9$$

Almost without effort, you've solved the equation, so you know that Jeremy donated 9 cans. With this information, you can go back to the chart, plug in 9 for j, and find out how many cans the other people donated: Liz donated 27, Alison donated 18, and Raymond donated 34. Finally, you can add up these numbers to conclude that the four people donated 88 cans altogether.

To check the numbers, read through the problem and make sure they work at every point in the story. For example, together Liz and Alison donated 45 cans, and Jeremy and Raymond donated 43, so the women really did donate 2 more cans than the men.

Crossing the finish line with five people

Here's one final example, the most difficult in this chapter, in which you have five people to work with.

> Five friends are keeping track of how many miles they run. So far this month, Mina has run 12 miles, Suzanne has run 3 more miles than Jake, and Kyle has run twice as far as Victor. But tomorrow, after they all complete a 5-mile run, Jake will have run as far as Mina and Victor combined, and the whole group will have run 174 miles. How far has each person run so far?

The most important point to notice in this problem is that there are two sets of numbers: the miles that all five people have run up to *today* and their mileage including *tomorrow.* And each person's mileage tomorrow will be 5 miles greater than his or her mileage today. Here's how to set up a chart:

	Today	***Tomorrow (Today + 5)***
Jake		
Kyle		
Mina		
Suzanne		
Victor		

With this chart, you're off to a good start to solve this problem. Next, look for that statement early in the problem that connects two people by either multiplication or division. Here it is:

> Kyle has run *twice as far* as Victor.

Because Victor has run fewer miles than Kyle, declare your variable as follows:

> Let v = the number of miles that Victor has run up to *today.*

Notice that I added the word *today* to the declaration to be very clear that I'm talking about Victor's miles *before* the 5-mile run tomorrow.

At this point, you can begin filling in the chart:

	Today	***Tomorrow (Today + 5)***
Jake		
Kyle	$2v$	$2v + 5$
Mina	12	17
Suzanne		
Victor	v	$v + 5$

As you can see, I left out the information about Jake and Suzanne because I can't represent it using the variable v. I've also begun to fill in the *Tomorrow* column by adding 5 to my numbers in the *Today* column.

Now I can move on to the next statement in the problem:

> But tomorrow ... Jake will have run as far as Mina and Victor combined... .

I can use this to fill in Jake's information:

	Today	***Tomorrow (Today + 5)***
Jake	$17 + v$	$17 + v + 5$
Kyle	$2v$	$2v + 5$
Mina	12	17
Suzanne		
Victor	v	$v + 5$

In this case, I first filled in Jake's *tomorrow* distance ($17 + v + 5$) and then subtracted 5 to find out his *today* distance. Now I can use the information that today Suzanne has run 3 more miles than Jake:

	Today	***Tomorrow (Today + 5)***
Jake	$17 + v$	$17 + v + 5$
Kyle	$2v$	$2v + 5$
Mina	12	17
Suzanne	$17 + v + 3$	$17 + v + 8$
Victor	v	$v + 5$

With the chart filled in like this, you can begin to set up your equation. First, set up a word equation, as follows:

> Jake tomorrow + Kyle tomorrow + Mina tomorrow + Suzanne tomorrow + Victor tomorrow = 174

Now just substitute information from the chart into this word equation to set up your equation:

$$17 + v + 5 + 2v + 5 + 17 + 17 + v + 8 + v + 5 = 174$$

As always, begin solving by isolating the algebraic terms:

$$v + 2v + v + v = 174 - 17 - 5 - 5 - 17 - 17 - 8 - 5$$

Next, combine like terms:

$$5v = 100$$

Finally, to get rid of the coefficient in the term $5v$, divide both sides by 5:

$$\frac{5v}{5} = \frac{100}{5}$$
$$v = 20$$

You now know that Victor's total distance up to *today* is 20 miles. With this information, you substitute 20 for v and fill in the chart, as follows:

	Today	***Tomorrow (Today + 5)***
Jake	37	42
Kyle	40	45
Mina	12	17
Suzanne	40	45
Victor	20	25

The *Today* column contains the answers to the question the problem asks. To check this solution, make sure that every statement in the problem is true. For example, tomorrow the five people will have run a total of 174 miles because

$$42 + 45 + 17 + 45 + 25 = 174$$

Copy down this problem, close the book, and work through it for practice.

Part VI

The Part of Tens

the part of tens

Math is full of important concepts. To discover ten of the most interesting, go to `www.dummies.com/extras/basicmathandprealgebra`.

In this part...

- Discover tricks to help you avoid making common mathematical mistakes
- Expand your understanding of math by learning how to distinguish between different kinds of numbers: natural numbers, integers, rational (and irrational) numbers, algebraic numbers, and more

Chapter 24

Ten Little Math Demons That Trip People Up

In This Chapter

- Knowing the multiplication table once and for all
- Understanding negative numbers
- Distinguishing factors and multiples
- Working confidently with fractions
- Seeing what algebra is really all about

The ten little math demons I cover in this chapter plague all sorts of otherwise smart, capable folks like you. The good news is that they're not as big and scary as you may think, and they can be dispelled more easily than you may have dared believe. Here, I present ten common math demons, with a short explanation to set them on a path away from you.

Knowing the Multiplication Table

A sketchy knowledge of multiplication can really hold back an otherwise good math student. Here's a quick quiz: the ten toughest problems from the multiplication table.

$8 \times 7 =$ ___ $9 \times 9 =$ ___

$7 \times 9 =$ ___ $6 \times 8 =$ ___

$6 \times 6 =$ ___ $8 \times 9 =$ ___

$7 \times 7 =$ ___ $9 \times 6 =$ ___

$8 \times 8 =$ ___ $7 \times 6 =$ ___

Can you do this, 10 for 10, in 20 seconds? If so, you're a multiplication whiz. If not, flip to Chapter 3 and work through my short, sweet, and simple program for nailing the multiplication table once and for all!

Adding and Subtracting Negative Numbers

It's easy to get confused when adding and subtracting negative numbers. To begin, think of adding a number as moving *up* and subtracting a number as moving *down.* For example:

$2 + 1 - 6$ means *up* 2, *up* 1, *down* 6

So if you go *up* 2 steps, then *up* 1 more step, and then *down* 6 steps, you've gone a total of 3 steps *down;* therefore, $2 + 1 - 6 = -3$.

Here's another example:

$-3 + 8 - 1$ means *down* 3, *up* 8, *down* 1

This time, go *down* 3 steps, then *up* 8 steps, and then *down* 1 step, you've gone a total of 4 steps *up;* therefore, $-3 + 8 - 1 = 4$.

You can turn every problem involving negative numbers into an up-and-down example. The way to do this is by combining adjacent signs:

- Combine a plus and minus as a *minus* sign.
- Combine two minus signs as a *plus* sign.

For example:

$-5 + (-3) - (-9)$

In this example, you see a plus sign and a minus sign together (between the 5 and the 3), which you can combine as a minus sign. You also see two minus signs (between the 3 and the 9), which you can combine as a plus sign:

$-5 - 3 + 9$ means *down* 5, *down* 3, *up* 9

This technique allows you use your up-and-down skills to solve the problem: *Down* 5 steps, then *down* 3 steps, and *up* 9 steps leaves you 1 step *up;* therefore, $-5 + (-3) - (-9) = 1$.

See Chapter 4 for more on adding and subtracting negative numbers.

Multiplying and Dividing Negative Numbers

When you multiply or divide a positive number by a negative number (or vice versa), the answer is always negative. For example:

$$2 \times (-4) = -8 \quad 14 \div (-7) = -2$$
$$-3 \times 5 = -15 \quad -20 \div 4 = -5$$

When you multiply two negative numbers, remember this simple rule: Two negatives always cancel each other out and equal a positive.

$$-8 \times (-3) = 24 \quad -30 \div (-5) = 6$$

For more on multiplying and dividing negative numbers, see Chapter 4.

Knowing the Difference between Factors and Multiples

Lots of students get factors and multiples confused because they're so similar. Both are related to the concept of divisibility. When you divide one number by another and the answer has no remainder, the first number is *divisible* by the second. For example:

$12 \div 3 = 4 \rightarrow$ 12 is *divisible* by 3

When you know that 12 is divisible by 3, you know two other things as well:

3 is a *factor* of 12 and 12 is a *multiple* of 3

In the positive numbers, the factor is always the *smaller* of the two numbers and the multiple is always the *larger.*

For more on factors and multiples, see Chapter 8.

Reducing Fractions to Lowest Terms

Math teachers usually request (or force!) their students to use the smallest-possible version of a fraction — that is, to reduce fractions to lowest terms.

To reduce a fraction, divide the *numerator* (top number) and *denominator* (bottom number) by a *common factor,* a number that they're both divisible by. For example, 50 and 100 are both divisible by 10, so

$$\frac{50}{100} = \frac{50 \div 10}{100 \div 10} = \frac{5}{10}$$

The resulting fraction, $\frac{5}{10}$, can still be further reduced, because both 5 and 10 are divisible by 5:

$$\frac{5}{10} = \frac{5 \div 5}{10 \div 5} = \frac{1}{2}$$

When you can no longer make the numerator and denominator smaller by dividing by a common factor, the result is a fraction that's reduced to lowest terms.

See Chapter 9 for more on reducing fractions.

Adding and Subtracting Fractions

Adding and subtracting fractions that have the same denominator is pretty simple: Perform the operation (adding or subtracting) on the two numerators and keep the denominators the same.

$$\frac{2}{7} + \frac{3}{7} = \frac{5}{7} \qquad \frac{8}{9} - \frac{7}{9} = \frac{1}{9}$$

When two fractions have different denominators, you can add or subtract them without finding a common denominator by using cross-multiplication, as shown here:

$$\text{To add: } \frac{3}{5} + \frac{1}{4} = \frac{(3 \times 4) + (5 \times 1)}{5 \times 4} = \frac{17}{20}$$

$$\text{To subtract: } \frac{2}{3} - \frac{1}{5} = \frac{(2 \times 5) - (3 \times 1)}{3 \times 5} = \frac{7}{15}$$

For more on adding and subtracting fractions, see Chapter 10.

Multiplying and Dividing Fractions

To multiply fractions, multiply their two numerators to get the numerator of the answer, and multiply their two denominators to get the denominator. For example:

$$\frac{3}{10} \times \frac{7}{8} = \frac{21}{80}$$

To divide two fractions, turn the problem into multiplication by taking the *reciprocal* of the second fraction — that is, by flipping it upside-down. For example:

$$\frac{2}{7} \div \frac{5}{6} = \frac{2}{7} \times \frac{6}{5}$$

Now multiply the two resulting fractions:

$$\frac{2}{7} \times \frac{6}{5} = \frac{12}{35}$$

For more on multiplying and dividing fractions, see Chapter 10.

Identifying Algebra's Main Goal: Find *x*

Everything in algebra is, ultimately, for one purpose: Find x (or whatever the variable happens to be). Algebra is really just a bunch of tools to help you do that. In Chapter 21, I give you these tools. Chapter 22 focuses on the goal of finding x. And in Chapter 23, you use algebra to solve word problems that would be much more difficult without algebra to help.

Knowing Algebra's Main Rule: Keep the Equation in Balance

The main idea of algebra is simply that an equation is like a balance scale: Provided that you do the same thing to both sides, the equation stays balanced. For example, consider the following equation:

$$8x - 12 = 5x + 9$$

To find x, you can do anything to this equation as long as you do it equally to both sides. For example:

Add 2: $8x - 12 = 5x + 9$ becomes $8x - 10 = 5x + 11$

Subtract $5x$: $8x - 12 = 5x + 9$ becomes $3x - 12 = 9$

Multiply by 10: $8x - 12 = 5x + 9$ becomes $80x - 120 = 50x + 90$

Each of these steps is valid. One, however, is more helpful than the others, as you see in the next section.

For more on algebra, see Chapters 21 through 23.

Seeing Algebra's Main Strategy: Isolate x

The best way to find x is to *isolate it* — that is, get x on one side of the equation with a number on the other side. To do this while keeping the equation balanced requires great cunning and finesse. Here's an example, using the equation from the preceding section:

Original problem	$8x - 12 = 5x + 9$
Subtract $5x$	$3x - 12 = 9$
Add 12	$3x = 21$
Divide by 3	$x = 7$

As you can see, the final step isolates x, giving you the solution: $x = 7$.

For more on algebra, see Chapters 21 through 23.

Chapter 25

Ten Important Number Sets to Know

In This Chapter

- Identifying counting numbers, integers, rational numbers, and real numbers
- Discovering imaginary and complex numbers
- Looking at how transfinite numbers represent higher levels of infinity

The more you find out about numbers, the stranger they become. When you're working with just the counting numbers and a few simple operations, numbers seem to develop a landscape all their own. The terrain of this landscape starts out uneventful, but as you introduce other sets, it soon turns surprising, shocking, and even mind blowing. In this chapter, I take you on a mind-expanding tour of ten sets of numbers.

I start with the familiar and comfy counting numbers. I continue with the integers (positive and negative counting numbers and 0), the rational numbers (integers and fractions), and real numbers (all numbers on the number line). I also take you on a few side routes along the way. The tour ends with the bizarre and almost unbelievable transfinite numbers. And in a way, the transfinite numbers bring you back to where you started: the counting numbers.

Each of these sets of numbers serves a different purpose, some familiar (such as accounting and carpentry), some scientific (such as electronics and physics), and a few purely mathematical. Enjoy the ride!

Counting on Counting (or Natural) Numbers

The *counting numbers* — also called the *natural numbers* — are probably the first numbers you ever encountered. They start with 1 and go up from there:

{1, 2, 3, 4, 5, 6, 7, 8, 9, 10, 11, 12, ...}

The three dots (or ellipsis) at the end tell you that the sequence of numbers goes on forever — in other words, it's infinite.

The counting numbers are useful for keeping track of tangible objects: stones, chickens, cars, cell phones — anything that you can touch and that you don't plan to cut into pieces.

The set of counting numbers is *closed* under both addition and multiplication. In other words, if you add or multiply any two counting numbers, the result is also a counting number. But the set isn't closed under subtraction or division. For example, if you subtract 2 – 3, you get –1, which is a negative number, not a counting number. And if you divide $2 \div 3$, you get $\frac{2}{3}$, which is a fraction.

If you place 0 in the set of counting numbers, you get the set of *whole numbers.*

Identifying Integers

The set of *integers* includes the counting numbers (see the preceding section), the negative counting numbers, and 0:

{..., –6, –5, –4, –3, –2, –1, 0, 1, 2, 3, 4, 5, 6, ...}

The dots, or ellipses, at the beginning and the end of the set tell you that the integers are infinite in both the positive and negative directions.

Because the integers include the negative numbers, you can use them to keep track of anything that can potentially involve debt. In today's culture, it's usually money. For example, if you have $100 in your checking account and you write a check for $120, you find that your new balance drops to –$20 (not counting any fees that the bank charges!).

The set of integers is *closed* under addition, subtraction, and multiplication. In other words, if you add, subtract, or multiply any two integers, the result is also an integer. But the set isn't closed under division. For example, if you divide the integer 2 by the integer 5, you get the fraction $\frac{2}{5}$, which isn't an integer.

Knowing the Rationale behind Rational Numbers

The *rational numbers* include the integers (see the preceding section) and all the fractions between the integers. Here, I list only the rational numbers from –1 to 1 whose denominators (bottom numbers) are positive numbers less than 5:

$$\left\{\ldots, -1\ldots, -\frac{3}{4}\ldots, -\frac{2}{3}\ldots, -\frac{1}{2}\ldots, -\frac{1}{3}\ldots, -\frac{1}{4}\ldots, 0\ldots, \frac{1}{4}\ldots, \frac{1}{3}\ldots, \frac{1}{2}\ldots, \frac{2}{3}\ldots, \frac{3}{4}\ldots, 1, \ldots\right\}$$

The ellipses tell you that between any pair of rational numbers are an infinite number of other rational numbers — a quality called the *infinite density* of rational numbers.

Rational numbers are commonly used for measurement in which precision is important. For example, a ruler wouldn't be much good if it measured length only to the nearest inch. Most rulers measure length to the nearest $\frac{1}{16}$ of an inch, which is close enough for most purposes. Similarly, measuring cups, scales, precision clocks, and thermometers that allow you to make measurements to a fraction of a unit also use rational numbers. (See Chapter 15 for more on units of measurement.)

The set of rational numbers is closed under the Big Four operations. In other words, if you add, subtract, multiply, or divide any two rational numbers, the result is always another rational number.

Making Sense of Irrational Numbers

In a sense, the irrational numbers are a sort of catchall; every number on the number line that isn't rational is irrational.

By definition, no *irrational number* can be represented as a fraction, nor can an irrational number be represented as either a terminating decimal or a repeating decimal (see Chapter 11 for more about these types of decimals). Instead, an irrational number can be approximated only as a *nonterminating, nonrepeating decimal:* The string of numbers after the decimal point goes on forever without creating a pattern.

The most famous example of an irrational number is π, which represents the circumference of a circle with a diameter of 1 unit. Another common irrational number is $\sqrt{2}$, which represents the diagonal distance across a square with a side of 1 unit. In fact, all square roots of nonsquare numbers (such as $\sqrt{3}$, $\sqrt{5}$, and so forth) are irrational numbers.

Irrational numbers fill out the spaces in the real number line. (The *real number line* is just the number line you're used to, but it's continuous; it has no gaps, so every point is paired with a number.) These numbers are used in many cases where you need not just a high level of precision, as with the rational numbers, but the *exact* value of a number that you can't represent as a fraction.

Irrational numbers come in two varieties: *algebraic numbers* and *transcendental numbers.* I discuss both types of numbers in the sections that follow.

Absorbing Algebraic Numbers

To understand *algebraic numbers,* you need a little information about polynomial equations. A *polynomial equation* is an algebraic equation that meets the following conditions:

- Its operations are limited to addition, subtraction, and multiplication. In other words, you don't have to divide by a variable.
- Its variables are raised only to positive, whole-number exponents.

You can find out more about polynomials in *Algebra For Dummies,* by Mary Jane Sterling (Wiley). Here are some polynomial equations:

$$2x + 14 = (x + 3)^2$$
$$2x^2 - 9x - 5 = 0$$

Every algebraic number shows up as the solution of at least one polynomial equation. For example, suppose you have the following equation:

$$x^2 = 2$$

You can solve this equation as $x = \sqrt{2}$. Thus, $\sqrt{2}$ is an algebraic number whose approximate value is 1.4142135623... (see Chapter 4 for more information on square roots).

Moving through Transcendental Numbers

A *transcendental number,* in contrast to an algebraic number (see the preceding section), is *never* the solution of a polynomial equation. Like the irrational numbers, transcendental numbers are a sort of catchall: Every number on the number line that isn't algebraic is transcendental.

The best-known transcendental number is π, whose approximate value is 3.1415926535.... Its uses begin in geometry but extend to virtually all areas of mathematics. (See Chapter 16 for more on π.)

Other important transcendental numbers come about when you study *trigonometry,* the math of right triangles. The values of trigonometric functions — such as sines, cosines, and tangents — are often transcendental numbers.

Another important transcendental number is *e,* whose approximate value is 2.718281828459.... The number *e* is the base of the natural logarithm, which you probably won't use until you get to pre-calculus or calculus. People use *e* to do problems on compound interest, population growth, radioactive decay, and the like.

Getting Grounded in Real Numbers

The set of *real numbers* is the set of all rational and irrational numbers (see the earlier sections). The real numbers comprise every point on the number line.

Like the rational numbers (see "Knowing the Rationale behind Rational Numbers," earlier in this chapter), the set of real numbers is closed under the Big Four operations. In other words, if you add, subtract, multiply, or divide any two real numbers, the result is always another real number.

Trying to Imagine Imaginary Numbers

An *imaginary number* is any real number multiplied by $\sqrt{-1}$.

To understand what's so strange about imaginary numbers, it helps to know a bit about square roots. The *square root* of a number is any value that, when multiplied by itself, gives you that number. For example, the square root of 9 is 3 because $3 \times 3 = 9$. And the square root of 9 is also –3 because $-3 \times -3 = 9$. (See Chapter 4 for more on square roots and multiplying negative numbers.)

The problem with finding $\sqrt{-1}$ is that it isn't on the real number line (because $\sqrt{-1}$ isn't in the set of real numbers). If it were on the real number line, it would be a positive number, a negative number, or 0. But when you multiply any positive number by itself, you get a positive number. And when you multiply any negative number by itself, you also get a positive number. Finally, when you multiply 0 by itself, you get 0.

If $\sqrt{-1}$ isn't on the real number line, where is it? That's a good question. For thousands of years, mathematicians believed that the square root of a negative number was simply meaningless. They banished it to the mathematical nonplace called *undefined,* which is the same place they kept fractions with a denominator of 0. In the 19th century, however, mathematicians began to find these numbers useful and found a way to incorporate them into the rest of math.

Mathematicians designated $\sqrt{-1}$ with the symbol i. Because it didn't fit onto the real number line, i got its own number line, which looks a lot like the real number line. Figure 25-1 shows some numbers that form the imaginary number line.

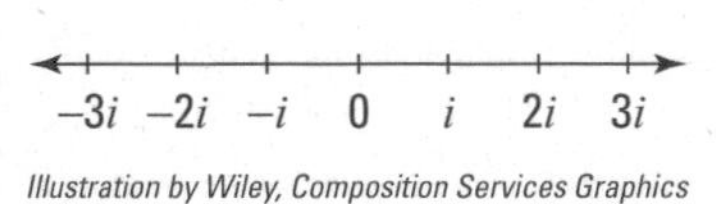

Illustration by Wiley, Composition Services Graphics

Figure 25-1: Numbers on the imaginary number line.

Even though these numbers are called imaginary, mathematicians today consider them no less real than the real numbers. And the scientific application of imaginary numbers to electronics and physics has verified that these numbers are more than just figments of someone's imagination.

Grasping the Complexity of Complex Numbers

A *complex number* is any real number (see "Getting Grounded in Real Numbers," earlier in this chapter) plus or minus an imaginary number (see the preceding section). Consider some examples:

$$1 + i \quad 5 - 2i \quad -100 + 10i$$

Getting inside subsets

Many sets of numbers actually fit inside other sets. Mathematicians call these nesting sets *subsets.* For instance, the set of integers is called $\mathbb{Z}$ for short. Because the set of counting or natural numbers (represented as $\mathbb{N}$) is completely contained within the set of integers, $\mathbb{N}$ is a subset, or part, of $\mathbb{Z}$.

The set of rational numbers is called $\mathbb{Q}$. Because the set of integers is completely contained within the set of rational numbers, $\mathbb{N}$ and $\mathbb{Z}$ are both subsets of $\mathbb{Q}$.

$\mathbb{R}$ stands for the set of real numbers. Because the set of rational numbers is completely contained within the set of real numbers, $\mathbb{N}$, $\mathbb{Z}$, and $\mathbb{Q}$ are all subsets of $\mathbb{R}$.

The set of complex numbers is called $\mathbb{C}$. Because the set of real numbers is completely contained within the set of complex numbers, $\mathbb{N}$, $\mathbb{Z}$, $\mathbb{Q}$, and $\mathbb{R}$ are all subsets of $\mathbb{C}$.

The symbol $\subset$ means "is a subset of" (see Chapter 20 for details on set notation). So here's how the sets fit inside each other:

$$\mathbb{N} \subset \mathbb{Z} \subset \mathbb{Q} \subset \mathbb{R} \subset \mathbb{C}$$

You can turn any real number into a complex number by just adding $0i$ (which equals 0):

$$3 = 3 + 0i \quad -12 = -12 + 0i \quad 3.14 = 3.14 + 0i$$

These examples show you that the real numbers are just a part of the larger set of complex numbers.

Like the rational numbers and real numbers (check out the sections earlier in this chapter), the set of complex numbers is closed under the Big Four operations. In other words, if you add, subtract, multiply, or divide any two complex numbers, the result is always another complex number.

Going beyond the Infinite with Transfinite Numbers

The *transfinite numbers* are a set of numbers representing different levels of infinity. Consider this for a moment: The counting numbers (1, 2, 3, ...) go on forever, so they're infinite. But there are *more* real numbers than counting numbers.

In fact, the real numbers are *infinitely more infinite* than the counting numbers. Mathematician Georg Cantor proved this fact. He also proved that, for every level of infinity, you can find another level that's even higher. He called these ever-increasing levels of infinity *transfinite,* because they transcend, or go beyond, what you think of as infinite.

The lowest transfinite number is $\aleph_0$ (aleph null), which equals the number of elements in the set of counting numbers ({1, 2, 3, 4, 5, ...}). Because the counting numbers are infinite, the familiar symbol for infinity (∞) and $\aleph_0$ mean the same thing.

The next transfinite number is $\aleph_1$ (aleph one), which equals the number of elements in the set of real numbers. This is a higher order of infinity than ∞.

The sets of integers, rational, and algebraic numbers all have $\aleph_0$ elements. And the sets of irrational, transcendental, imaginary, and complex numbers all have $\aleph_1$ elements.

Higher levels of infinity exist, too. Here's the set of transfinite numbers:

$$\{\aleph_0, \aleph_1, \aleph_2, \aleph_3, ...\}$$

The ellipsis tells you that the sequence of transfinite numbers goes on forever — in other words, that it's infinite. As you can see, on the surface, the transfinite numbers look similar to the counting numbers (in the first section of this chapter). That is, the set of transfinite numbers has $\aleph_0$ elements.

Index

• Symbols and Numerics •

• A •

• B •

• C •

• D •

• E •

• F •

• G •

• H •

• I •

• K •

• L •

• M •

• N •

• O •

• P •

• Q •

• R •

• S •

• V •

• W •

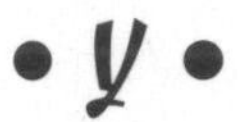

• Z •

About the Author

Mark Zegarelli is a math and test prep teacher, as well as the author of eight *For Dummies* (Wiley) books, including *SAT Math For Dummies, ACT Math For Dummies,* and *Calculus II For Dummies*. He holds degrees in both English and math from Rutgers University and lives in Long Branch, New Jersey, and San Francisco, California.

Dedication

I dedicate this book to the memory of my mother, Sally Ann Zegarelli (Joan Bernice Hanley).

Author's Acknowledgments

Writing this second edition of *Basic Math & Pre-Algebra For Dummies* was an entirely enjoyable experience, thanks to the support and guidance of Lindsay Lefevere of John Wiley & Sons, Inc., editors Tracy Barr and Krista Hansing, and technical reviewer Mike McAsey. And many thanks to my assistant, Chris Mark, for his unfailing diligence and enthusiasm.

And thanks to the great folks at Borderlands Café, on Valencia Street in San Francisco, for creating a quiet and friendly place to work.

Finally, a shout out to my awesome nephews, Jake and Ben.

Publisher's Acknowledgments

Executive Editor: Lindsay Sandman Lefevere

Project Editor: Tracy L. Barr

Copy Editor: Krista Hansing

Technical Editor: Michael McAsey

Project Coordinator: Sheree Montgomery

Cover Image: ©Jiri Moucka illustrations/Alamy

Apple & Mac

iPad For Dummies, 5th Edition
978-1-118-49823-1

iPhone 5 For Dummies, 6th Edition
978-1-118-35201-4

MacBook For Dummies, 4th Edition
978-1-118-20920-2

OS X Mountain Lion For Dummies
978-1-118-39418-2

Blogging & Social Media

Facebook For Dummies, 4th Edition
978-1-118-09562-1

Mom Blogging For Dummies
978-1-118-03843-7

Pinterest For Dummies
978-1-118-32800-2

WordPress For Dummies, 5th Edition
978-1-118-38318-6

Business

Commodities For Dummies, 2nd Edition
978-1-118-01687-9

Investing For Dummies, 6th Edition
978-0-470-90545-6

Personal Finance For Dummies, 7th Edition
978-1-118-11785-9

QuickBooks 2013 For Dummies
978-1-118-35641-8

Small Business Marketing Kit For Dummies, 3rd Edition
978-1-118-31183-7

Careers

Job Interviews For Dummies, 4th Edition
978-1-118-11290-8

Job Searching with Social Media For Dummies
978-0-470-93072-4

Personal Branding For Dummies
978-1-118-11792-7

Resumes For Dummies, 6th Edition
978-0-470-87361-8

Success as a Mediator For Dummies
978-1-118-07862-4

Diet & Nutrition

Belly Fat Diet For Dummies
978-1-118-34585-6

Eating Clean For Dummies
978-1-118-00013-7

Nutrition For Dummies, 5th Edition
978-0-470-93231-5

Digital Photography

Digital Photography For Dummies, 7th Edition
978-1-118-09203-3

Digital SLR Cameras & Photography For Dummies, 4th Edition
978-1-118-14489-3

Photoshop Elements 11 For Dummies
978-1-118-40821-6

Gardening

Herb Gardening For Dummies, 2nd Edition
978-0-470-61778-6

Vegetable Gardening For Dummies, 2nd Edition
978-0-470-49870-5

Health

Anti-Inflammation Diet For Dummies
978-1-118-02381-5

Diabetes For Dummies, 3rd Edition
978-0-470-27086-8

Living Paleo For Dummies
978-1-118-29405-5

Hobbies

Beekeeping For Dummies
978-0-470-43065-1

eBay For Dummies, 7th Edition
978-1-118-09806-6

Raising Chickens For Dummies
978-0-470-46544-8

Wine For Dummies, 5th Edition
978-1-118-28872-6

Writing Young Adult Fiction For Dummies
978-0-470-94954-2

Language & Foreign Language

500 Spanish Verbs For Dummies
978-1-118-02382-2

English Grammar For Dummies, 2nd Edition
978-0-470-54664-2

French All-in One For Dummies
978-1-118-22815-9

German Essentials For Dummies
978-1-118-18422-6

Italian For Dummies 2nd Edition
978-1-118-00465-4

Available in print and e-book formats.

Available wherever books are sold. For more information or to order direct: U.S. customers visit www.Dummies.com or call 1-877-762-2974. U.K. customers visit www.Wileyeurope.com or call (0) 1243 843291. Canadian customers visit www.Wiley.ca or call 1-800-567-4797.

Connect with us online at www.facebook.com/fordummies or @fordummies